The Earth's Climate:
Past and Future

This is Volume 29 in
INTERNATIONAL GEOPHYSICS SERIES
A series of monographs and textbooks
Edited by WILLIAM L. DONN

A complete list of the books in this series appears at the end of this volume.

The Earth's Climate: Past and Future

M. I. Budyko

State Hydrological Institute
Leningrad, USSR

Translated by the Author

1982

ACADEMIC PRESS
A Subsidiary of Harcourt Brace Jovanovich, Publishers
New York   London
Paris   San Diego   San Francisco   São Paulo   Sydney   Tokyo   Toronto

ACADEMIC PRESS, INC.
111 Fifth Avenue, New York, New York 10003

United Kingdom Edition published by
ACADEMIC PRESS, INC. (LONDON) LTD.
24/28 Oval Road, London NW1   7DX

Library of Congress Cataloging in Publication Data

Budyko, M. I. (Mikhail Ivanovich)
    The earth's climate: Past and future

    (International geophysics series)
    Includes index.
    1. Climatic changes. 2. Man--Influence on nature.
I. Title: Earth's climate. II. Series.
QC981.8.C5B83   551.6   81-17673
ISBN 0-12-139460-3   AACR2

PRINTED IN THE UNITED STATES OF AMERICA

82 83 84 85   9 8 7 6 5 4 3 2 1

THE EARTH'S CLIMATE: PAST AND FUTURE. Translated from the
original Russian edition entitled КЛИМАТ В ПРОШЛОМ
И БУДУЩЕМ, published by ГИДРОМЕТЕОРОЛОГИЧЕСКОЕ
ИЗДАТЕЛЬСТВО, ЛЕНИНГРАД, 1980.

"That's the effect of living backwards," the Queen said kindly: "it always makes one a little giddy at first—"

"Living backwards!" Alice repeated in great astonishment. "I never heard of such a thing!"

"—but there's one great advantage in it, that one's memory works both ways."

"I'm sure *mine* only works one way," Alice remarked. "I ca'n't remember things before they happen."

"It's a poor sort of memory that only works backwards," the Queen remarked.

Lewis Carroll
"Through the Looking-Glass"

Contents

Preface

Since 1961 the author, with his collaborators, has studied the problem of changes in the global climate due to natural factors and man's activities. The results of the studies were set forth in several publications, including the monograph "Climatic Change," published in 1974. Since that time, many new results have been obtained, which have not yet been tied together and presented in a monograph. These results support the previously drawn conclusion that a drastic change in global climate toward warming will occur during the next few decades if modern trends for generating energy continue. Whereas several years ago there were only a few proponents of this idea, by now their number has greatly increased.

Conclusions about anthropogenic climatic change were also supported in the final documents of several international scientific conferences held in the late 1970s. This drew the attention of the World Meteorological Organization and other scientific organizations and government institutions in different countries.

There is now a widespread belief that it may be possible to change the global climate in the near future. Therefore, the author thought it appropriate to present a description of studies of the above problem in this book, giving only a brief account of the questions discussed in the previous monograph, and focusing on the results obtained during the past few years.

Leningrad, 1979

Some minor changes have been introduced into the English edition of the book.

Leningrad, 1981

Chapter

1

Introduction

1.1 THE PROBLEM OF FUTURE CLIMATIC CONDITIONS

Man and climate

Climatic conditions have influenced man's life and activities, beginning with the first stages of his existence. Human evolution seems to have been considerably affected by the Pliocene and Pleistocene climatic changes that resulted from global cooling that was most intensive in high latitudes. This caused an increase in the mean meridional air temperature gradient. As a consequence, the system of atmospheric circulation changed, the high pressure belts being moved to lower latitudes.

Due to insufficient precipitation within these belts, the moisture conditions changed over vast tropical areas, leading to the replacement of tropical forests in several regions by savannas and semideserts.

There is reason to believe that before that climatic change, man's ancestors inhabited tropical forests in Africa, their way of life being similar to that of other higher primates. In open areas, where man's ancestors

1

could not seek shelter in trees, they had virtually no protection against representatives of the numerous predators of the African savanna. In addition, the disappearance of many edible plants restricted their vegetable diet. The ability to walk upright, which completely freed man's hands, and the more developed brain helped man's ancestors to survive. The extreme conditions of rigid natural selection seemed to lead to a faster rate of evolution, particularly with regard to the nervous system and mental ability.

Contemporary man emerged in the epoch of climatic changes caused by the last Quaternary glaciation. Drastic changes in ecological conditions seemed to sharpen the struggle for existence and to aid in the development of the species *Homo sapiens* (see, for example, Gerasimov, 1970; Budyko, 1977a).

Although primitive people settled only where there were suitable climatic conditions, one should note their ability to protect themselves against unfavorable weather conditions, even in the early stages of their existence. In the Paleolithic age, man used fire not only for cooking but to protect himself from the cold. With the advent of clothes and dwellings, people could inhabit new territories with colder climates. As a result, tens of thousands of years ago, man settled on all of the continents except Antarctica. Nevertheless, in spite of the great technological progress of mankind, the density of population in different areas is still strongly dependent on climatic conditions.

Currently, vast areas of polar deserts—Antarctica, central Greenland, and others—have no permanent population. The population density is very low in large areas of arid regions; the most arid zones are not populated at all. The density of population is also very low in the Arctic and sub-Arctic, in many mountainous regions, and in a number of areas with a predominance of swamps, etc.

Therefore, the greater part of the world population is concentrated in a limited region where the climatic conditions are most favorable for human life and activities. The size of this region is gradually increasing, but even with the modern population explosion, the development of regions with less favorable climatic conditions is comparatively slow.

It is known that the efficiency of man's economic activities is greatly affected by climate. The various forms of agriculture in different climatic zones are essentially dependent on meteorological conditions. Agricultural productivity within any zone can vary considerably because of differences in the meteorological regimes in individual regions. Everywhere, agricultural productivity fluctuates to a greater or lesser degree depending on variations in weather and climate. Crop yield is particularly variable in regions of unstable moistening. Variations in moistening regime in these regions significantly affect the global food balance because of their large contribution to total crop production.

In the past few years, due to rapid growth of population, food production in many countries has begun to lag behind the level necessary for answering minimum needs. Under these conditions, any climatic fluctuations that decrease yields can have serious consequences. Knowledge of climatic changes is of great importance in agricultural planning because the effectiveness of modern agrotechnology depends upon reliable information on future meteorological regimes.

This information is also necessary for planning hydraulic-engineering projects and heating systems within cold climate zones, and solving many other problems of great economic importance.

Until recently, information on climates of the recent past has been sufficient to answer questions on future climatic conditions. Now, due to rapidly increasing demands for more accurate information on the future climate from various sectors of the national economy, this method is quite often inadequate.

Recent studies indicate that during the last decades, man's activities have begun to affect the global climate. The effects are mainly due to:

(1) increasing CO_2 concentration in the atmosphere,
(2) increasing production of energy for use by man, and
(3) changing aerosol concentration in the atmosphere.

The most important effect on climate can be attributed to the rise in concentration of atmospheric CO_2 caused by the burning of ever-increasing amounts of coal, oil, and other fossil fuels and by some other forms of economic activity. Because CO_2 is almost transparent to shortwave solar radiation and absorbs a large fraction of longwave radiation, it appears to be one of the factors causing the greenhouse effect in the atmosphere, increasing the temperature of the lower air layer. For this reason an increase in the amount of CO_2 should lead to global warming.

The second influence on climate is associated with the fact that energy consumption by various kinds of human activities results in an additional heating of the atmosphere. The energy consumed by man is converted into heat, the greater portion of which is an additional energy source causing an elevation of atmospheric temperature.

Among the significant contemporary sources of the energy man consumes, only hydroenergy and the energy contained in wood and agricultural products are part of the transformed energy of solar radiation annually absorbed by the earth. The expenditure of these kinds of energy does not change the earth's heat budget and does not cause additional heating. However, these kinds of energy are a small part of the total amount consumed by man. The energy obtained from the burning of coal, oil, and natural gas, and nuclear energy are sources of heat that are not related to the transformation of solar radiation energy in the present epoch. With further growth in energy consumption, the heat produced by man can constitute a considerable part of the energy of solar radiation that is absorbed at the earth's surface, causing warming.

Atmospheric aerosol, whose particles scatter and absorb short- and longwave radiation, exerts a more complex effect on climatic conditions. As the mass of atmospheric aerosol grows due to man's activities, the radiation regime of the atmosphere changes. In particular cases this can bring about both cooling and warming.

Since global climate is affected by modern man's activities, the problem of studying climatic changes is one of the most important in atmospheric science. This is true, not only because many economic activities depend on climatic conditions, but also because global climatic changes influence almost all the natural processes in the biosphere.

It is known that many natural processes that form the geographical environment are affected by man's activities. This anthropogenic effect appears all over the world. In particular, flora and fauna on our planet have been greatly changed. Many species of animals have been annihilated by man; the populations of even more species have decreased sharply, threatening them with extinction. Among these are almost all large land animals, many species of marine animals, including marine mammals, which have particularly suffered from man's activities.

Land vegetation over a great part of the continents has undergone enormous changes. Over vast areas, natural vegetation has been destroyed to produce arable land. The remaining forests are mostly secondary, i.e., strongly modified by man's impact as compared to natural vegetation. Many steppe and savanna regions have been appreciably changed due to intensive cattle pasturing.

The process of soil formation in several regions has been affected by man's impact on natural vegetation cover. This has resulted in changing the physical and chemical properties of the soils. The arable land soils have changed greatly due to systematic cultivation, fertilization, and the removal of a considerable amount of plant biomass.

Man's impact on the hydrological regime on land is rapidly increasing. The runoff on not only small but many large rivers is essentially changed because of its regulation by means of hydrotechnical constructions. A considerable amount of the river runoff is removed to meet industrial and domestic demands and to irrigate arable lands. Large artificial reservoirs with areas comparable with those of large natural lakes drastically change the regimes of evaporation and runoff over vast territories.

It should be mentioned that the global change in the natural environment is generally the sum of local effects on natural processes. All these changes have assumed planetary proportions, not because man has changed natural global processes, but because the local (or regional) effects have spread over vast territories.

There is another situation in which man influences the global climate. Anthropogenic climatic change has an effect on the regime of rivers and lakes, vegetation and animals, soil, and other components of the natural environment all over the planet, with the result that human activity has become, for the first time, the main factor in the evolution of the biosphere as a whole.

It is obvious that the far reaching consequences of this impact should be studied.

The goal of the study

The practical significance of the question of future climatic conditions depends on the scale of probable changes in climate and the time of their occurrence. If these changes are sufficiently great and occur in the near future, the problem of predicting expected climatic changes becomes urgent.

This problem is important primarily because the economies of all countries depend heavily on contemporary climate. Noticeable climatic changes will require enormous investments for adjusting the economy to the new climatic conditions.

The period of time for which it is necessary to have information on climatic changes seems to be comparable with the expected useful lifetime of climate-dependent industrial and agricultural constructions. For the most durable of these, this period is 50–100 years. The possibility of climate modification should be taken into account when planning such construction.

There is another criterion for assessing the duration of the period over which it is desirable to have information on possible climatic changes. This is the time necessary to plan and implement some measures to control climatic changes and to adjust the economy for these changes. For these steps to be carried out, many complex scientific and technical problems must be solved; therefore this time should be no less than several decades. Thus it is desirable to have information on possible climatic changes over the period of about 50 to 100 years.

The problem of predicting anthropogenic climatic changes differs considerably from that of weather forecasting. In working out the latter, we need to analyze only the physical processes in the atmosphere and hydrosphere, while in studying the former, we must consider, in addition, temporal variations in man's activities.

In this connection, the problem of forecasting climatic changes consists of two major elements: the prediction of progress in some aspects of economics (the growth of fuel consumption that increases CO_2 concentration in the atmosphere, the expansion of energy production, etc.) and the calculation of those climatic changes that would take place with corresponding changes in economics.

This leads to two important characteristics of the forecasts indicated. First, they will inevitably be conditional. Human activities are not independent of man's impact on climatic conditions. If an economic activity might lead to largely unfavorable changes in the global climate, then the character of the economic activity would probably change before these variations occurred. Therefore, the goal of a climatologist is not to forecast the real climate of the future, but to compute the parameters of such a climate, taking into account several possible versions of economic progress ("scenarios"). Considering the results obtained, it is possible to optimize the long-term planning of the economy, taking measures against unfavorable climatic changes. Hence, the prediction of possible climatic changes is the basis for climatic control in the future.

The second characteristic of forecasting the future climate concerns its potential accuracy. Since the quantitative forecasting of economic progress for several decades is difficult for many reasons, the accuracy of such a forecast cannot be high. Therefore, an approximate method, which can correctly assess the sign and order of magnitude of possible climatic changes, would seem to be sufficient when predicting the future climate. These estimates are of practical significance because they make it possible to single out those versions of economic progress which may cause large (i.e., most important for a national economy) climatic changes that exceed probable calculation errors.

Since estimates of future climatic conditions should be used for long-term economic planning, we must try to achieve the highest reliability. However, it is difficult to obtain really reliable estimates of future climates, in particular because of the imperfection of the climatic theories that are used to compute the various anthropogenic effects on climatic conditions.

To overcome this difficulty, this study uses two independent methods for forecasting the future climate.

One of these uses data on past climates. This method is based on the study of the evolution of atmospheric chemical composition during the geological past. The results of this study are presented in Chapter 2. The second method consists in the utilization of a semiempirical theory of the atmospheric thermal regime; the results are discussed in Chapter 3. The results obtained by using this theory to explain the physical mechanism of natural climatic changes occurring during the geological past and at present are presented in Chapter 4. Good agreement between the calculated results of natural climatic changes and empirical data indicates that the theory given can be used for estimating future climatic conditions. The impact of man's economy on modern climate is evaluated in Chapter 5. Conclusions about the future climate drawn from both of the above methods are set forth in Chapter 6. These conclusions seem to be reliable; there is good agreement between the results obtained by using both of the methods.

1.2 THE ADVANCEMENT OF STUDIES

Causes of climatic change

To answer questions about future climatic conditions, it is of great importance to determine the causes of natural climatic changes.

Although many investigators have been interested in this question, until now there has been no generally accepted opinion on the physical mechanism of climatic change for the modern epoch and the geological past. Because of the absence of an accurate climatic theory and the lack of special observational data, the different hypotheses about the causes of climatic changes were not widely accepted and they aroused many objections. Considerable progress in this field has been made, particularly during the last decade. As a result, we now understand the physical mechanisms of climatic changes much better.

This brief review deals mainly with those previous studies of climatic changes that have proved to be the most fruitful in developing the present theories.

In a number of studies of the causes of climatic change, emphasis is placed on the climatic effect of CO_2 concentration. Over a hundred years ago Tyndall pointed out that atmospheric CO_2, together with water vapor, absorbed longwave radiation in the atmosphere. Variations in CO_2 concentration could result in climatic fluctuations (Tyndall, 1861).

Later Arrhenius (1896, 1903) and Chamberlin (1897, 1898, 1899) studied the question of the climatic effect of atmospheric CO_2. They supposed that variations in CO_2 concentration could have caused the Quaternary glaciations.

Arrhenius studied the absorption of radiation fluxes in the atmosphere.

He suggested a numerical model for determining the temperature near the earth's surface, depending on properties of the atmosphere. Using this model Arrhenius found that a 2.5- to 3-fold increase in CO_2 concentration raises the air temperature by 8–9°C and a 38–45% decrease lowers the temperature by 4–5°C.

Considering the materials of geological investigations, Arrhenius stated that the mass of CO_2 in the present atmosphere is only a small fraction of that which existed in the past, much of which was absorbed from the atmosphere by the formation of carbon-containing rocks. In this connection, Arrhenius concluded that the CO_2 concentration of the atmosphere could vary over a wide range. These variations, he believed, exerted a considerable influence on air temperature, sufficient for glaciations to arise or to disappear.

The geological aspect of this problem was studied by Chamberlin. Considering the CO_2 balance, Chamberlin stated that the amount of CO_2 coming from the lithosphere varied noticeably depending on the level of volcanic activity and other factors.

The expenditure of CO_2 in geological processes also changed appreciably, in particular in accordance with the surface area of the rocks exposed to atmospheric erosion. As the area increased, the expenditure of CO_2 by weathering grew as well.

Chamberlin assumed that the glaciations resulted from the intensive process of orogenesis and elevation of continents, which led to an increase in the erosion basis and the surface area of weathered rocks and to a decrease in atmospheric CO_2 concentration. To corroborate this assumption, Chamberlin carried out some calculations that could not, however, be considered a complete quantitative model of the process.

It was later established that within the $13–17$-μm CO_2 radiation absorption band, an overlap occurs with part of the absorption band of H_2O. This decreases the influence of CO_2 concentration variations on the thermal regime.

Considering this effect, Callender (1938) obtained smaller values than those of Arrhenius for the temperature variations at the earth's surface. Using Callender's data, a doubling of the CO_2 concentration increases the air temperature by 1.5°, and the effects of variations of CO_2 concentration on temperature decrease with increasing concentration.

Möller (1963) has studied the problem of the effect of CO_2 on the atmospheric thermal regime. He determined that when air temperature varies, the absolute humidity changes, whereas the relative humidity is more or less constant. An increase in the absolute humidity as the temperature rises increases the absorption of longwave radiation in the atmosphere, further raising the temperature. Möller found that the temperature increase near the earth's surface caused by doubling the CO_2 concentration with constant absolute humidity is 1.5°, whereas with constant relative humidity this value is several times larger. Möller also mentioned that the effect of CO_2 on the thermal regime could be compensated for by comparatively small variations in the absolute humidity or cloudiness.

Thorough investigation of the dependence of the air temperature on the CO_2 concentration of the atmosphere was carried out by Manabe and Wetherald (1967). In their work they noted the inaccuracy of calculations by Möller, who estimated variations in the thermal regime of the atmosphere using only data on the heat balance of the earth's surface without considering the atmospheric heat balance as a whole.

Manabe and Wetherald computed the vertical temperature distribution in the atmosphere, taking into account the absorption of longwave radiation by water vapor, carbon dioxide, and ozone. They used a vertical distribution of the relative humidity taken from empirical data. It was assumed that the temperature distribution was determined by conditions of local radiation balance if the lapse rate did not exceed 6.5° km^{-1}. This value was considered a maximum because of the limiting effect of convection on the growth of the vertical gradient.

Manabe and Wetherald disclosed that under average cloud conditions with constant relative humidity, doubling the CO_2 concentration raised the temperature at the earth's surface by 2.4°. Reduction to half the present CO_2 level lowered the air temperature by 2.3°.

In a later study, Manabe and Wetherald (1975) computed the CO_2 effect on climate using a three-dimensional model of atmospheric general circulation, taking into account water exchange, heat exchange in the atmosphere, in the oceans, and on the continents (with idealized topography), and the feedback between snow and ice cover and the atmospheric thermal regime. They found that doubling the CO_2 concentration raises the mean air temperature at the earth's surface by 2.9°. Similar estimates have been obtained in a number of recent studies that used various simplified models of climate theory.

The relationship between the atmospheric thermal regime and the level of CO_2 concentration is of importance in understanding climatic changes. The assumption of the author that a decrease in atmospheric CO_2 concentration exerted an essential influence on climatic cooling in the late Cenozoic era (Budyko, 1974) was supported by data on the evolution of the atmospheric chemical composition in the Phanerozoic (Budyko, 1977a; Budyko and Ronov, 1979). From these data, it was concluded that variations in the CO_2 concentration caused the lowering of the mean air temperature that began in the late Mesozoic era and continued in the Cenozoic era.

Studies of the climatic fluctuations that occurred in the Pleistocene did not support the assumption by Arrhenius and Chamberlin of the dependence of these fluctuations on variations in atmospheric chemical composition, although their basic idea, that the evolution of atmospheric chemical composition is important in climatic changes, was highly fruitful.

A French mathematician (Adhèmar, 1842) was the first to elucidate the causes of the Quaternary glaciations and to point out that changes in the earth's surface position relative to the sun could result in glacier advance. This concept was elaborated by Kroll and others and was treated comprehensively by Milankovich (1920, 1930, 1941).

To explain climatic changes during the Quaternary period, Milankovich used the calculated results of the secular change of three astronomical elements: the earth's orbital eccentricity, the inclination of the earth's axis to the ecliptic plane, and the time of precession of the equinoxes due to the precession of the earth's axis of rotation (i.e., the moments during the year when the sun crosses the celestial equator and the distance between the sun and the earth is the shortest). All these elements vary with time because of the effect of the moon and the other planets on the earth's movement: the period of change in the eccentricity is ~92,000 yr; in the inclination of the axis, ~40,000 yr; and in the precession of the equinoxes, ~21,000 yr.

Although oscillations in these elements do not influence the amount of solar radiation coming to the outer boundary of the atmosphere of the earth as a whole, they do alter the total radiation received by various latitudinal zones in different seasons of the year.

Milankovich worked out an approximate theory of radiation and the thermal regime of the atmosphere, from which it followed that the surface air temperature in the warm period of the year, at middle and high latitudes, can vary by several degrees, depending on the relationships among the above astronomical elements.

Such a variation usually coincided, to a greater or lesser degree, with a temperature change of opposite sign in the cold season. However, as is known from empirical investigations, the glaciation regime depends mainly on the temperature of the warm season (when ice melts). The thermal conditions of the cold season exert far less influence on ice cover.

By using the calculated results of the astronomical elements for the Quaternary period, Milankovich assumed that the periods of temperature decrease during the warm season in the zone where most glaciation developed (60°–70° latitude) corresponded to the glacial epochs, and the periods of temperature rise to the intervals between them.

Köppen was the first to compare Milankovich's results and climatic conditions during the Quaternary period. He found good agreement between Milankovich's conclusions and the history of Quaternary glaciations (Köppen and Wegener, 1924). Similar comparisons were carried out by other authors but their conclusions proved to be contradictory. For this reason, the question of the climatic effect of astronomical factors has generated lively discussion.

To determine the causes of the Quaternary climatic changes, one can use semiempirical theories of the atmospheric thermal regime (Budyko, 1968a; Sellers, 1969) by which it was possible to quantitatively take into account the effect on the thermal regime of air-temperature–polar-ice feedback.

During discussions of the question of the causes of the Quaternary glaciations, several attempts were made to construct numerical models of the development of glaciations which would make it possible to relate changes in the area of polar ice cover to the changing position of the earth's surface in relation to the sun.

In a number of cases these attempts led to the conclusion that the influence of astronomical factors upon the development of glaciations was unimportant. This was due to the fact that the models used did not adequately take into account the ice-cover–atmospheric-thermal-regime feedback. When this feedback was considered, it was shown that changes in the position of the earth's surface in relation to the sun exerted an essential influence on the development of the Quaternary glaciations (Budyko and Vasishcheva, 1971; Budyko, 1972; Berger, 1973). In these works, semiempirical models were used for calculating the atmospheric thermal regime. Later, a similar conclusion was obtained using more general models of climate theory (Suarez and Held, 1976).

Of great importance for determining the causes of the Quaternary glaciations was an empirical investigation which established a close agreement between the intervals of decreasing total summer radiation at high latitudes and an improved chronology of Quaternary glaciations (Hays *et al.*, 1976).

Studies of the atmospheric aerosol effect on climatic conditions are of great importance for understanding the physical mechanism of present-day climatic changes.

In the 18th century, Franklin was the first to assume that variations in atmospheric transparency affect climate.

In the first half of the 20th century, Humphreys (1913, 1929) and others advanced the hypothesis that volcanic eruptions contributed to present-day climatic changes. Humphreys stated that clouds of small particles, which he believed to consist of volcanic dust, appeared in the stratosphere after explosive volcanic eruptions. He estimated their effect on the radiation regime and came to the conclusion that they could noticeably weaken the flux of shortwave radiation coming to the earth's surface while scarcely changing the longwave emission reflected into space. Humphreys calculated that the amount of particles in the atmosphere of the Northern Hemisphere after the Katmai eruption in 1912 was sufficient to lower the mean surface air temperature by several degrees if the particles remained in the atmosphere over a fairly long period. Humphreys found from observational data that the air temperature, averaged over vast areas, actually decreased by several tenths of a degree after intense volcanic eruptions. He believed that this comparatively small temperature change after eruptions could be due to a short residence time of volcanic dust particles in the atmosphere.

This idea was later corroborated and advanced by Wexler (1953, 1956) who assumed that the smallest dust particles could remain in the stratosphere for several years after volcanic eruptions. Wexler thought that the 1920–1930 warming could be attributed to a gradual clearing from the atmosphere of the volcanic dust produced by several explosive eruptions in the early 20th century.

Brooks (1950) attached much importance to the climatic effects of volcanism. Using data on volcanic rocks, he composed a table of the intensity of volcanic activity from the beginning of the Paleozoic up to the present time. Brooks believed that an increase in volcanism was one of the factors causing the advance of continental glaciations.

Lamb's paper (1970) is one of the recent studies devoted to investigating the climatic effect of volcanism. It presents information on all explosive volcanic eruptions occurring after 1500 that considerably attenuated atmospheric transparency for shortwave radiation. Lamb, as well as Humphreys and Wexler, considered volcanic eruptions to be one of the major factors in present-day climatic changes.

The secular variation of mean air temperature in the Northern Hemisphere was calculated from data on solar radiation variations using a semiempirical theory of the thermal regime. This calculation showed that the main cause of contemporary natural climatic variations was a change in the concentration of aerosol particles in the lower stratosphere (Budyko, 1969a). This conclusion was later drawn in a number of other studies (Pollack *et al.*, 1975; Oliver, 1976; and others).

In this paragraph are stated the results of a study of the physical mechanism of climatic changes, at present and during the last part of the Cenozoic era when the shapes of the continents did not differ considerably from their modern ones. In the remote past (discussed in Chapter 4) the climate was to some degree influenced by variations in the structure of the earth's surface.

As seen from the brief review given here, the major causes of climatic changes during the last tens of millions of years and in the present epoch have been to a considerable extent determined.

The question of the causes of natural climatic changes is of great importance for studying man's impact upon climate. For this reason, Chapter 4 is devoted to a comprehensive consideration of these changes.

Anthropogenic changes of climate

Callender (1938) was the first to propose that an increase in atmospheric CO_2 concentration due to modern man's economic activities would result in global warming. Although not all the conclusions of this study proved to be correct, many of them were significantly ahead of the science of that time.

Taking into account the amount of fossil fuel used in the preceding 50 years, Callender concluded that 75% of the CO_2 produced remained in the atmosphere. Not only at that time but much later it was thought that almost all the CO_2 produced by man dissolved in the ocean.

Callender was also the first to evaluate the relationship between air temperature and CO_2 mass variation, taking into account the influence of water vapor on the atmospheric radiation regime. In this calculation he obtained the effect of "saturation," i.e., a decrease in the rate of change of air temperature with an increase in the amount of CO_2. Callender estimated the value of the mean air temperature rise with doubling of the CO_2 concentration by means of a rather simple computation. Although this value seemed to be underestimated, it agrees well with several results obtained in similar studies many years later that were based on fairly complicated theoretical models.

It is worth noting that Callender tried to estimate the climatic change caused by man-made CO_2 up to the year 2200. He believed that the burning of existing fossil fuel resources would lead to a tenfold increase in atmospheric CO_2 mass (this differs but slightly from some modern estimates). Assuming the CO_2 production rate to be constant during the period in question, Callender concluded that by 2200 the mean air temperature would rise approximately $0.6°$. This would result in an ~ 130-km displacement of climatic zones to higher latitudes.

Although Callender did not take into account that the rate of temperature increase thus computed ought to be largely understated due to a rapid increase in the annual fossil fuel consumption, his attempt to estimate future climatic conditions by considering the effect of anthropogenic factors is of considerable interest.

It should be noted that Callender compared the warming that occurred in the late 19th and early 20th centuries with the effect of CO_2 mass growth on air temperature. He calculated the mean rate of warming from data observed at a comparatively small number of stations and considered the possible influence of a local temperature rise in cities on the results obtained. The calculated value proved to be noticeably underestimated. His conclusion that warming depends on CO_2 mass growth appeared to be imperfectly substantiated.

It is interesting to study the comments of outstanding English meteorologists of that time (Simpson, Brent, Brooks, and others), published along with Callender's article. It seems that the participants in the discussion did not fully understand the importance of the results obtained by Callender. This could be explained by the fact that Callender was not a meteorologist (he was engaged in the field of the technical sciences).

In the USSR the problem of anthropogenic changes in global climate was first discussed at the scientific conferences held in 1961–1962 (see the reviews Gal'tzov, 1961; Gal'tzov and Chaplygina, 1962; and the papers Budyko, 1961, 1962a,b).

As seen from these, at that time data were obtained which enabled us to state that with further progress in energetics the global climate would change essentially in the near future. These works underlined a particular significance, for climatic changes, of the stability of polar sea ice, because its area is related by positive feedback to the air temperature in high latitudes. Due to this feedback, a comparatively small initial temperature elevation caused by man's activities could lead to partial or complete melting of polar sea ice, which ought to be accompanied by a drastic climatic change in the high and mid-latitudes of the Northern Hemisphere. Since 1961 the USSR Hydrometeorological Service has been studying the anthropogenic change in global climate and, in 1975, to broaden these investigations, the Department for Climatic Change Research was established at the State Hydrological Institute. During the 1960–1970s in the USSR, the problem of anthropogenic climatic change was discussed at many scientific conferences held by the Hydrometeorological Service (at present, Goskomgidromet—the USSR State Committee for Hydrometeorology and Control of Natural Environment) and the USSR Academy of Sciences.

Since the prospect of changes in the global climate is of great importance for all countries of the world, international scientific cooperation should play a considerable role in the study of this problem.

The first US–USSR conference devoted to this problem was held in Los Angeles in 1966; at it the question of the stability of polar sea ice under the conditions of anthropogenic climatic change received primary emphasis (Proceedings of Symposium on the Arctic Heat Budget and Atmospheric Circulation, 1966). In the 1970s, several Soviet–American symposia on anthropogenic climate change (Tashkent in 1976, Leningrad in 1977, Dushanbe in 1978, Tbilisi in 1979) and palaeoclimatology (in the USSR in 1976, in the USA in 1978) were held. Leading scholars in the field of climate research from both countries took part in these symposia.

Since 1970, and particularly since 1975, interest in the question of possible anthropogenic climatic change has greatly increased in different countries. In the USA, in 1970, a conference on the problem of man's impact on the environment was held, with the results published as a monograph (SCEP, 1970). One of the conclusions of the monograph was the following: ". . . we stress that the long-term potential consequences of CO_2 effects on the climate or of societal reaction to such threats are so serious that much more must be learned about future trends of climate change."

In this monograph it has been assumed that until the end of the 20th century, the probability of an appreciable change in climate caused by economic activity is low. Later the problem of anthropogenic climatic change was discussed at the conference held in 1971 in Stockholm at which the monograph "Inadvertent Climate Modification" (SMIC, 1971) was prepared. This monograph was mainly devoted to discussing the physical mechanism of man's impact on climate and to determining the problems of further investigations in this field. Although this monograph contained some new ideas (as compared with SCEP), there were no definite conclusions on the character of future anthropogenic changes of climate or the time of their occurrence. Both monographs cited the conclusion drawn by Manabe (1970) that by 2000 the mean air temperature could be 0.8° higher than in 1900, due to CO_2 mass growth. However, this result has not been used for assessing future climatic conditions.

Manabe's conclusion was not regarded as a climatic forecast because the climate theories of that time were not believed to be sufficiently reliable to be used in calculating climatic changes. In particular, many investigators thought that even a small change in cloud cover due to temperature variation could compensate for any anthropogenic climate modification. Concurrent with this, it was widely held that a variation of a few tenths of a degree in the mean planetary air temperature is of no great practical importance.

The author held another viewpoint. In a work (Budyko, 1972) published soon after the monographs mentioned above, a semiempirical theory of the atmospheric thermal regime was used to compute natural climatic changes that occurred during the Quaternary period and the last 100 years. Since the results obtained agreed well with observational data, it was concluded that the theory mentioned can be utilized to forecast anthropogenic climatic changes. In this paper a graph was constructed showing probable temperature variation due to man's impact for the next 100 years. It can be seen from this graph that until the year 2050 the climate will be mainly affected by an increase in atmospheric CO_2 mass, contributing to a rise in the mean air temperature near the earth's surface. Afterward, the climate could be affected by an increase in the consumption of energy that is not obtained by the transformation of solar energy reaching the earth's surface at present. This factor would accelerate the process of warming.

Approximate computations made in this paper demonstrated that warming could lead, in 2000, to the recession of the polar sea ice boundary in the Arctic by 2° of latitute and, in 2050, to its complete melting, the estimates of ice melting rate being seemingly underestimated. It was also concluded that even a comparatively small warming would bring about an increase in drought frequency in a number of regions on the continents, a decrease in river runoff, and a fall in the levels of many lakes and interior seas. To prevent the unfavorable consequences of a climatic change toward warming, a method based on increasing the aerosol concentration in the lower layers of the stratosphere was suggested as a means of influencing the atmospheric processes.

The conclusion of the paper was that, although it would be premature to use its results for economic planning, they could be employed for assessing the time available for obtaining more accurate information about future trends of climatic conditions. It was stated that this time should not exceed ten years, otherwise it would be difficult to prevent harmful effects of climatic changes upon the national economy.

The viewpoint that warming would develop in the near future was expressed when scientists believed that the mean air temperature was decreasing, dropping by 0.3° in the Northern Hemisphere for the period from the late 1930s to the mid-1960s.

In the mid-1970s it was established that in the middle or late 1960s cooling ceased and a warming process began. This increased the urgency of studying the problem of the CO_2 effect on climate.

In 1976–1979 several conclusions about the possible consequences of CO_2 mass growth were published (World Meteorological Organization, 1976, 1977) in which a considerable change in climatic conditions as a result of the burning of ever growing amounts of coal, oil, and other kinds of fuel was considered possible. The introduction to the proceedings of the conference on global chemical cycles and their alterations by man states:

According to most projections, the CO_2 concentration will increase by 30–40% over the preindustrial level in 2000 and will approximately double by 2025. There is some controversy as to the level which can be tolerated without serious climatic consequences, but it has been suggested that a doubling or perhaps even a lesser increase of the CO_2 content of the atmosphere may already lead to unacceptable climatic consequences (Stumm, 1977).

In 1977 in the USA a commission of the National Academy of Sciences issued the report "Energy and Climate." In this report the influence of anthropogenic factors on climate (including aerosol, gas ejections, and heat production) was discussed (National Academy of Sciences, 1977).

The introduction to the report begins:

Worldwide industrial civilization may face a major decision over the next few decades—whether to continue reliance on fossil fuels as principal sources of energy or to invest the research and engineering effort, and the capital that will make it possible to substitute other energy sources for fossil fuels within the next 50 years. The second alternative presents many difficulties, but the possible climatic consequences of reliance on fossil fuels for another one or two centuries may be so severe as to leave no other choice.

As indicated in this report, the climate during subsequent centuries should be mainly affected by the growth of CO_2 concentration in the atmosphere. This could lead to an increase of more than 6° in the mean air temperature near the earth's surface, with an increase several times greater in high latitudes.

Variations in climate and atmospheric composition are of particular importance for agriculture because they can result in an increase in the photosynthesis rate, transference of agroclimatic regions to higher latitudes, shifts of the boundaries of arid zones, and other effects.

The authors, after indicating several possible ways of preventing anthropogenic changes in climate, proposed that the most realistic method for solving this problem is to reduce the consumption of fossil fuels. The report of the National Academy of Sciences was widely discussed in the press. In a number of cases the authors made statements in addition to the conclusions of the report. For example, the chairman of the Commission on Energy and Climate said that we must end the use of fossil fuels by 2050. He suggested that it would be necessary to spend from 20 to 100 million dollars per year for the next 50 years to study this problem (Revelle, 1977).

The question of future trends in anthropogenic climatic changes was discussed in 1978 at the US–USSR Symposium on the CO_2 problem. The concluding document of the symposium indicated the possibility of a considerable change in global climate, due to CO_2 mass growth in the atmosphere, that would essentially affect natural conditions and the economy.

The First World Conference on Climate was held by the World Meteorological Organization and other international institutions in Geneva in 1979. Among the participants in the conference were several Soviet specialists in the field of geography and geophysics: I. P. Gerasimov, E. K. Fedorov, Yu. A. Israel, and G. I. Marchuk. In his report, Fedorov (1979) said:

Future climatic changes are inevitable. They would become apparent and probably irreversible during the near several decades. . . . In this connection it is evident that some strategy

should be worked out, i.e. the system of beforehand planned actions which could help mankind to avoid negative consequences of expected climatic changes. . . . What are the main
elements of this strategy? The first and major is naturally the forecasting of climatic
change. . . . The second is the assessment of those consequences to which those or these
natural or man-made climate changes can lead. . . . The third is the working out of recommendations on such actions which could help to avoid the negative consequences of climate
changes or to avoid the changes themselves.

Gerasimov (1979), Fedorov (1979), Israel (1979) and Marchuk (1979)
have substantiated major sections of the USSR climate research program,
which is aimed at assessing, in particular, present-day climatic changes
and to forecast their future tendencies.

The conference accepted a declaration that appealed to all the countries
of the world, stating, in particular, that the burning of fossil fuel, the destruction of forests, and soil cultivation have increased the CO_2 concentration in the atmosphere by 15% over the past hundred years, an increase
that is continuing at present. Therefore, the development of a warming
trend, which would be most significant in middle and high latitudes, is expected. Since man-made climatic changes would be favorable in some regions and unfavorable in others, complex social and technological problems of international significance emerge. To solve these problems,
unprecedented efforts are necessary to organize studies of the problems
of climatic changes and to create new forms of international cooperation.

In addition to the statements made at a number of scientific conferences
and by organizations, the views of many scientists concerning the question of anthropogenic climatic change due to the growth of CO_2 concentration have been recently published.

Broecker (1975) has pointed out that in the very near future, due to the
growth of CO_2 concentration, natural cooling would cease and the change
to a quickly developing global warming trend would result in considerable
modification of water exchange that would affect agriculture in different
countries.

A similar conclusion was drawn in an investigation by Bolin (1975). In
the conclusion of Bolin's "Energy and Climate," he says:

. . . But the very possibility that significant changes may take place within a time period of
less than 100 years, maybe within 50 years, is an exceedingly important fact. Within another
decade or two reality will most likely show how relevant these fears are. Even now due
regard to such possibilities must be paid when planning our future energy supply, above all
by retaining the utmost degree of flexibility.

A book prepared at Oak Ridge National Laboratory begins with the
words: "Most scientists viewing the accelerated burning of fossil fuels
now agree that excess CO_2 will warm the earth's surface temperature significantly" (Baes et al., 1976).

Flohn (1977b) in his report at the Dahlem Conference gave a forecast of probable climatic changes during the next hundred years due to CO_2 concentration growth in the atmosphere. In the introduction to the first issue of a new international scientific journal, *Climatic Change,* Flohn (1977a) also emphasized that the danger of large-scale climatic change could occur before the economy has been prepared for this change. In this connection he compared the climatologists of the present with Cassandra, the heroine of a Greek myth. (A daughter of Priam, King of Troy, she was endowed with the gift of prophecy but fated by Apollo never to be believed.) Everybody was sceptical of the danger that she predicted, which resulted in the destruction of the city. At the end of his article Flohn said "It is our generation which bears responsibility for a global scale problem facing our grandchildren—let us take care to match it."

At a meeting of a committee of the US Congress in May 1976, S. Schneider said:

> Professor Bolin has already said that by the year 2000 perhaps the climate could warm as much as 1° Celsius. I have argued that that can be serious. We do not know whether it would be through melting of the ice caps or not.
>
> We also expect that the climate in high latitudes will respond much more strongly than that of lower latitudes to such a warming, so a 1 degree global average increase in temperature might mean a 5 degree increase in the poles. What that does is change the equator-to-pole temperature difference, which is the driving force behind the weather. That might move, for example, the grain belt several hundred miles north.
>
> Is this a global catastrophe? . . . If this occurs at a time when global margins (of food production) are already extremely tight, the potential seriousness of that situation, to me, deserves the word "catastrophe" [Schneider (1976, p. 82)].

In addition to this, Schneider pointed out ". . . there will always be someone who will argue that we do not know precisely what will happen —and they will always be right, because science is not a certain game. . . . The question is, how much proof is enough, and how much chance do you want to take?" (Schneider, 1976).

Schneider also mentioned that the accuracy of calculating the future climate from the growth of CO_2 concentration would be greatly improved by the end of our century, but by that time it could prove to be too late to take the necessary measures for reconstructing energy schemes. Many investigators favored similar views (Landsberg, 1976, 1977; Olson *et al.,* 1978).

In 1976–1978, the U.S. Congress considered a bill to organize a study of this problem. The attention of the Congress to this question is explained by the great importance of present-day climatic changes. This is considered in a number of publications by American scientists, among them the book "The Genesis Strategy" by Schneider and Mesirow (1976). The title of the book is taken from the first part of the Bible (the Book of Genesis) in which the story of Joseph is set forth. He foretold the coming of seven years of plentiful harvests and then seven years of famine and he advised the Egyptian Pharaoh to lay away the plentiful harvests against the years of scarcity. This book, which was written for the general reader, offers a wealth of information about the complex workings of climatic change and a minute examination of the relationships between climatic change and problems in ecology, economics, and politics.

The law on the climatic program adopted in the United States in 1978 emphasizes the great importance of climatic changes in problems related to food and energy production. The law states that it is necessary to improve methods of forecasting climatic changes, the necessary prerequisites being available for this. In accordance with this law, complex scientific researches are being organized to work out methods for predicting natural and man-made climatic changes that can be used for long-term economic planning.

The number of scientific publications devoted to the problem of climatic change is one of the criteria of increased attention to this problem. Since the mid-1970s, the number of such publications per year has grown considerably as compared with the 1960s. The publication of an international scientific journal *Climatic Change,* edited by S. Schneider, has recently begun.

Extended investigations of this problem in other countries started later than in the USSR. Therefore, in many cases, the authors obtained results similar to those obtained in the USSR. A number of aspects of research on climatic changes have been studied extensively in the USSR. Among them are the use of simplified theoretical models of climate in calculating its changes, studies of the evolution of atmospheric chemical composition, and the utilization of palaeoclimatic data for estimating the future climate. Some problems have been studied principally in other countries (for example, the question of the CO_2 balance in the modern epoch).

Although speedy but not always well-organized development of research on the problems of climatic change has frequently led to more or less considerable disagreement in solving individual questions, an ever-growing correlation is observed in the principal conclusions drawn about present-day climatic changes by leading specialists of different countries. Among these is the conclusion that a large global climatic change will occur if we continue our present reliance on an energy program based on the burning of fossil fuels.

The first maps have recently been constructed of the climate that will exist as as result of an appreciable CO_2 concentration growth. They are based both on calculations using climatic theories and on palaeoclimatic data, similar results being obtained with the use of these two independent methods for estimating the future climate.

The available information on future climate is comprehensively discussed in Chapter 6. This information supports the conclusion that the conditions of the future climate should be drastically different from those of the contemporary climate. All the components of the natural environment and man's economic activity will be greatly affected by anthropogenic changes in global climate. Therefore, the question of future climatic change is the most urgent problem of contemporary atmospheric science.

Chapter

2

Evolution
of the Atmosphere

2.1 THE ATMOSPHERE

General information

The earth's atmosphere consists mainly of nitrogen and oxygen, which in dry air make up 78 and 21% of the volume, respectively. Of the minor constant atmospheric constituents, carbon dioxide and ozone considerably influence the physical state of the atmosphere and biological processes.

The atmosphere also contains water vapor, the quantity of which varies over a wide range. It is generally present in concentrations of 0.1–1.0% and occasionally reaches as much as a few percent of the atmospheric volume. The mean water vapor content is about 2.4 g cm^{-2}.

The relative amounts of nitrogen, oxygen, and carbon dioxide vary little with altitude in the troposphere and the lower stratosphere. The water vapor concentration, in contrast, decreases rapidly from the earth's surface to the upper boundary of the troposphere. Stratospheric water vapor concentration is low and relatively independent of height. The relative amount of ozone in the troposphere is not high and it reaches a maximum in the lower stratosphere where it is formed by a photochemical reaction depending on solar ultraviolet radiation.

The total mass of the atmosphere is $\sim 5.3 \times 10^{21}$ g (i.e., less than 10^{-6} of the earth's mass). The atmospheric pressure at sea level is ~ 1033 g cm^{-1}.

Atmospheric density and pressure decrease with altitude. Pressure decreases to half its surface value at a height of about 5 km.

The troposphere, the lowest layer of the atmosphere, extends up to 8–10 km in polar regions and 16–18 km in the tropics. In the troposphere, the air temperature generally decreases with height at an average rate of $6°$ km^{-1}. The stratosphere extends above the troposphere. The temperature in the stratosphere increases up to a height of about 50 km. More than two-thirds of the atmosphere's mass is concentrated in the troposphere.

Physical processes in the troposphere determine weather fluctuations and have a basic influence on the climatic conditions of our planet. These processes include absorption of solar radiation, formation of longwave radiation flux (which varies little in the higher atmospheric layers), the general circulation of the atmosphere, and the hydrological cycle involved in the formation of clouds and the initiation of precipitation.

Since the atmosphere is more or less transparent to shortwave radiation, most of the solar flux that the earth absorbs is received at the earth's surface, which appears to be the main source of heat for the atmosphere. Heat energy is transferred from the earth's surface to the atmosphere by longwave radiation and turbulent heat exchange. The heat consumed by evaporation from land and from water reservoirs is also released into the atmosphere in the process of condensation of the water vapor.

Because the earth is a sphere, most of the solar energy is absorbed in the low latitudes. The surface air temperature in this belt is generally higher than in the middle and high latitudes. The temperature difference appears to be the main source of energy for atmospheric and oceanic circulation. Atmospheric circulation is an intricate system of horizontal and vertical motions of the air. In the stratosphere, vertical motions are generally weak in contrast to the troposphere, where they are quite significant. The general circulation of the atmosphere and the oceans ensures that a considerable amount of heat will be transferred from low to high and middle latitudes, leading to a decrease in the meridional gradient of the air temperature.

Heating of the earth's surface by solar energy results in upward air fluxes that cause cloud formation and precipitation when the atmospheric air becomes saturated.

Evaporation from the ocean surface appears to supply most of the water vapor to the atmosphere. In addition, the atmosphere over the continents receives water vapor by evaporation from the land surface. This local evaporation is negligible in areas with low precipitation.

Of some importance in weather and climate are processes occurring in the stratosphere that are associated in particular with the dissipation and absorption of radiation fluxes by liquid and solid particles of atmospheric aerosols. These particles—ranging from ~ 10 Å to ~ 10 μm—can be observed both in the troposphere and in the upper layers of the atmosphere. Weathering of the earth's surface and evaporation of sea-water droplets are the major sources of the atmospheric aerosols. Aerosols are also added to the atmosphere when solid particles are ejected by volcanic eruptions and they are formed from the sulfurous gas that is released into the atmosphere during volcanic eruptions and other natural processes. Some of the aerosols enter the upper atmosphere from outer space. In addition to the natural aerosol sources, many aerosol particles are of anthropogenic origin, resulting from industrial effluents, soil erosion, forest fires, and other sources. Aerosol concentration decreases rapidly with altitude but it may increase in certain layers (aerosol layers).

The atmosphere profoundly affects biological processes on land and in the water. The atmospheric constituents most important in biological processes are the oxygen consumed by respiration of organisms and mineralization of organic matter, the carbon dioxide used in the photosynthesis of autotrophic plants, and the ozone that absorbs some of the potentially harmful ultraviolet radiation of the sun. No complex living things can exist on the planet without the atmosphere. The moon, which has no atmosphere and no life, presents a fine example of this.

The evolution of the atmosphere is closely related to geological and geochemical processes as well as to the activity of living organisms.

Numerous theories exist that deal with the initial chemical composition of the atmosphere but they are difficult to verify using the available data. It is possible that water vapor, carbon dioxide, and nitrogen entered the primitive atmosphere from the depths of the earth's crust mainly through volcanic eruptions during the early stages of the earth's history.

A comparatively high concentration of CO_2 in the ancient atmosphere probably had a great effect on the climate, even if the total atmospheric mass was far less than it is presently. This factor intensified the greenhouse effect and perhaps compensated for the reduced heat income from the sun, producing relatively high air temperatures (see Chapter 4).

Since the evolution of the autotrophic plants (probably about 3 billion years ago), O_2 started accumulating in the atmosphere, first at a slow rate, then more rapidly after vegetation cover developed on the continents in the Palaeozoic. Fluctuations in atmospheric composition and changes in its mass resulted from irregular volcanic activity in the geological past. This question is thoroughly considered below.

The atmosphere has also greatly influenced the evolution of the lithosphere because it was a powerful factor in the physical weathering that formed the surface of the continents during the entire history of the lithosphere. Atmospheric precipitation and winds carrying fine fragments of rocks long distances contributed to this process. Temperature fluctuations and other atmospheric factors were also important in rock weathering. Chemical weathering, particularly the effects of atmospheric O_2 and water on the rocks, was even more important in the evolution of the lithosphere.

The evolution of the atmosphere largely determined the hydrosphere's development, since the water balance of the reservoirs depends directly on the precipitation and evaporation regimes. On the other hand, the atmospheric processes were under the influence of the state of the hydrosphere, particularly that of the ocean. Generally, the development of the atmosphere and the hydrosphere was an integrated process.

Atmospheric factors predestined to a great extent the evolution of living organisms. More or less complex organisms could emerge after the O_2 content of the atmosphere was high enough for the development of the aerobic processes that supplied the necessary energy income for these organisms.

The evolution of autotrophic plants was closely related to the atmospheric CO_2 content. Changes in CO_2 concentration influenced the structure of the vegetation cover and hence all the organisms whose vital activity was associated with autotrophic plants.

Contemporary climate

Climatic conditions reflect the physical state of the atmosphere in different areas of the globe. This state underwent dramatic changes in the past and is still varying.

The climatic conditions of the last century have been determined by the instrumental meteorological observations carried out by the world network of climatic stations that was formed in the second half of the nineteenth century.

The observational data have shown that the elements of the meteorological regime vary noticeably with time. In addition to their periodic fluctuations (diurnal or annual variation), there are nonperiodic fluctuations within different time scales. For short periods (of the order of days or

Table 2.1
Air Temperature and Precipitation

Latitude (degrees north)	90–80	80–70	70–60	60–50	50–40	40–30	30–20	20–10	10–0
January temperature (°C)	−31	−25	−22	−10	−1	11	19	25	27
July temperature (°C)	−1	2	12	14	20	26	28	28	27
Precipitation (cm yr⁻¹)	19	26	52	80	75	77	73	114	201
Latitude (degrees south)	0–10	10–20	20–30	30–40	40–50	50–60	60–70	70–80	80–90
January temperature (°C)	27	26	25	20	12	5	0	−8	−13
July temperature (°C)	26	24	18	14	8	1	−12	−30	−42
Precipitation (cm yr⁻¹)	150	122	91	103	108	101	67	25	11

months), nonperiodic variations in the meteorological regime correspond to weather fluctuations. These inhomogeneous spatial variations are mainly the result of instability in the atmospheric circulation.

In addition to irregular fluctuations in the elements of the meteorological regime, long-term variations frequently reveal themselves (for periods as long as a few years and more). They are of the same nature over vast areas and reflect climatic changes.

Since contemporary climatic fluctuations are relatively small, one can use mean values of the meteorological elements over a period of several decades to describe the present climate. Such averaging makes it possible to eliminate the effect of the instability of the atmospheric circulation on the elements of the meteorological regime.

We present here a brief synopsis of the contemporary climate, emphasizing two meteorological elements: the air temperature near the earth's surface and the total precipitation. Table 2.1 shows data on the mean latitudinal temperature for January and July as well as the annual mean latitudinal atmospheric precipitation reaching the earth's surface. The table shows that the average latitudinal air temperature near the earth's surface varies from a maximum near the equator to a minimum near the South Pole, the difference being almost 70°.

The origin of the variations in the mean latitudinal temperature is considered in detail in the next chapter. As mentioned above, temperature distribution is affected to a large extent by the spherical shape of the earth, which is responsible for the latitudinal change of the total radiation incident on the top of the atmosphere.

In the high latitudes where the air temperature rarely rises above the melting point permanent ice cover exists almost all the year round. Most of the Arctic ice cover consists of relatively thin sea ice; the Antarctic ice cover includes a thick continental glaciation covering nearly all of the Antarctic continent.

The polar sea ice of the Northern Hemisphere resembles a huge lens, which reaches a mean thickness of 3–4 m in the center, gradually diminishing toward the edges. In the Southern Hemisphere the sea ice has the shape of a broad ring encircling the Antarctic continent. The ice thickness there decreases with distance from the pole. Since sea ice consists of a great many separate ice fields, it is constantly moving under the influence of atmospheric and oceanic currents, resulting in glacial marginal changes.

Sea ice increases in extent during the cold seasons and diminishes in summer and in early fall. Ice cover, which has a high albedo, exerts a strong influence on climate. It lowers the air temperature in high latitudes, thereby increasing the meridional temperature gradient.

The mean air temperature near the earth's surface also varies considerably with longitude, primarily in accordance with the location of the continents and oceans.

The ocean heat regime affects a great portion of the continental surfaces, where a marine type of climate, with a relatively small annual amplitude in the air temperature, can be observed. In some extratropical areas of the continents, the effects of the oceanic thermal regime seem to be less noticeable. Here, the annual temperature amplitudes are high, corresponding to the conditions of a continental climate.

The distribution of the mean latitudinal precipitation reveals a major maximum in the equatorial belt, a rainfall decrease in the subtropics, two secondary maxima in the mid-latitudes, and a decrease in precipitation with increasing latitude in the polar areas.

Variations in the average latitudinal precipitation depend on the distribution of the mean air temperatures and especially on atmospheric circulation.

If other conditions remain the same, including relative humidity, the total rainfall will increase as the temperature rises because the amount of water vapor that can be used in the process of condensation also increases. If precipitation were completely independent of other factors, the mean latitudinal distribution would result in a single maximum at low latitudes.

An important factor in the initiation of rainfall is a vertical air velocity regime that affects the process of water vapor condensation, resulting in the formation of clouds and precipitation.

The general atmospheric circulation is closely related to the geographical distribution of stable pressure zones, the most important being the equatorial belt of low pressure, the high-pressure zone in tropical and subtropical latitudes, and the mid-latitudinal belt of frequent cyclonic formations.

Because downward air motions dominate in the high-pressure zone, the precipitation decreases considerably, producing two minima in the curve of latitudinal rainfall distribution. In the equatorial latitudes and in a few mid-latitudinal areas a high intensity of upward air motion increases the amount of precipitation.

The largest deserts of the world are situated in the high-pressure subtropical zone, where precipitation is negligible. The total rainfall also decreases in the interior continental areas in middle latitudes. In these regions, air currents carry little water vapor from the ocean, making relative humidity lower and water vapor condensation less intensive.

Therefore, the continental-humid climate covers mostly equatorial latitudes and marine climate areas in high and mid-latitudes, whereas low humidity conditions prevail in the high-tropical and subtropical latitudes as well as in the continental climate zone.

Climate has a profound effect on the geographical environment. Climate determines the water regime on land, defined mainly by river runoff, which is equal to the difference between precipitation and evaporation from the surface of the rivers' basins under average annual conditions. Evaporation depends largely on precipitation, radiation income, and thermal conditions, and to a lesser degree on the earth's surface structure. The influence of these climatic factors on river runoff is apparently dominating.

There is a high correlation between climatic conditions and the physical regime of the ocean. If the regime of ocean currents depends on wind, radiation, and thermal conditions over the ocean surface, then heat transfer by the ocean currents, in turn, is an essential climate-forming factor. In zones of warm and cold sea currents, which in some cases cover large areas, climatic conditions are drastically different.

Natural vegetation cover on land is profoundly dependent on climatic conditions. Thermal and radiation regimes, as well as moisture conditions, are the most vital of the meteorological factors that determine the development of plants. The relation between the animal world on land and climatic factors can be either direct (since there are climatic zones favorable for the existence of every species of animal) or determined by an animal's dependence on the natural vegetation, which is influenced by climatic factors.

Climatic parameters such as the radiation balance of the earth's surface, the thermal regime, and the total precipitation determine the location of geographic zones that reflect regularities of distribution of natural vegetation cover, the animal world, and soils, as well as river water regimes, the intensity of surface geomorphological processes, etc.

As a result of the essential dependence of various natural processes on climatic conditions, the natural environment all over the world usually varies with global climatic change.

2.2 THE CARBON CYCLE

Carbon dioxide

Carbon dioxide (CO_2) is a colorless gas which turns into a solid resembling snow (Dry Ice) at a temperature of $-78.5°C$.

The present atmosphere contains a relatively small amount of CO_2 (2.3×10^{18} g or 0.033% of the air volume). About 130×10^{18} g of CO_2 are dissolved in the oceans and other water reservoirs, roughly amounting to 0.01% of the hydrosphere's mass. Carbon dioxide is exchanged continually between the atmosphere and the hydrosphere by means of molecular and turbulent diffusion.

Measurements of the atmospheric CO_2 concentration have shown small variations in different geographical areas. The concentration also varies slightly with altitude in the troposphere and the lower stratosphere. A relatively constant CO_2 concentration in the atmosphere (as compared to that of water vapor) is explained by the smaller relative variability of CO_2 sources and sinks over the earth's surface. From observational data it follows that the CO_2 concentration increases slightly near the equator and decreases by $\sim 0.005\%$ in high latitudes. This occurs because CO_2 is more soluble in the cold ocean water of high latitudes than in warm tropical waters. As a consequence of this, the high-latitude atmosphere is lacking the portion of CO_2 that has dissolved in the ocean. The excess CO_2 is transferred by cold, deep currents to low latitudes, where it enters the atmosphere. This mechanism accounts for approximately 2×10^{16} g yr^{-1} of the CO_2 flux between the equator and the North Pole (Bolin and Keeling, 1963).

In addition to the CO_2 cycle between the atmosphere and the ocean, there is exchange of CO_2 among the atmosphere–hydrosphere system, living organisms, and the lithosphere. This circulation is one of the major mechanisms that maintain life on our planet.

A relatively small atmospheric and hydrospheric mass of CO_2 is of vital importance for living organisms. We know that CO_2 is assimilated by autotrophic plants through photosynthesis. The overwhelming majority of plants and animals exist by using energy that results from the activity of photoautotrophic plants.

In some cases CO_2 can also affect the animal world directly. Although the usual atmospheric concentration of CO_2 is harmless to aerobic animals, an increase of the concentration to a few percent or more can lead to the poisoning of many aerobic organisms. In particular, the mass deaths of animals in some areas of volcanic activity have confirmed this fact. Poisoning occurs when the release of an abundance of CO_2 from the depths of the lithosphere coincides with weak air circulation, hindering the dissipation CO_2. Carbon dioxide is less poisonous for soil aerobic animals, which generally exist in an atmosphere with high CO_2 concentrations.

It was mentioned in Chapter 1 that in addition to its great importance for living organisms, CO_2 affects the climate profoundly. These questions are addressed in the next chapters of the book.

Photosynthesis and oxidation of organic matter

Atmospheric CO_2 is removed from and released into the atmosphere through a few natural processes. A considerable amount of CO_2 dissolves annually in ocean waters and returns to the atmosphere from the oceans. This CO_2 cycle introduces little change in the total quantity of CO_2 in the atmosphere, since the CO_2 input is almost equal to its output.

An important component of atmospheric CO_2 consumption appears to be its assimilation by green plants through photosynthesis. The total photosynthetic reaction may be written in simplified form as

$$CO_2 + H_2O = CH_2O + O_2$$

This means that CO_2 consumption by synthesis of organic material is accompanied by a release of O_2.

In the course of photosynthesis, land plants use mostly CO_2 from the atmosphere and a much smaller amount of CO_2 from the soil air. Aquatic plants absorb the CO_2 dissolved in the hydrosphere. A considerable part of the organic matter produced by photosynthesis is used in the course of a plant's respiration and released by the atrophy of its individual parts. This leads to the restoration of the CO_2 used in the process of photosynthesis. As a result of the expenditure of a part of the organic matter, the productivity of autotrophic plants and consequently the total CO_2 consumption generally makes up half or two-thirds of the CO_2 used in photosynthesis.

Some investigators have tried to evaluate the total productivity of autotrophic plants in different geographical areas. The observational data show that productivity essentially varies according to the nature of the vegetation as well as the climatic and other external factors. With mounting observations, it was possible to draw maps of mean annual productivity (Lieth, 1964–1965; Rodin and Bazilevich, 1965).

Efimova (1977) constructed a map of world plant productivity on the continents based on an empirical relationship of productivity with precipitation and the radiation balance of the earth's surface. An evaluation from this study of the productivity of natural vegetation for the different continents is shown in Table 2.2.

It has been established that the highest productivity of terrestrial plants occurs under the conditions of a humid, warm climate, while in cold, high-latitude zones and in the driest areas of deserts productivity approaches zero. In the ocean the greatest amount of organic matter is produced by plankton at a depth of up to hundreds of meters. Lack of essential mineral compounds in ocean water often strongly limits the productivity of plants in the open sea. Because of this, productivity in shallow reservoirs and in inshore regions is generally higher than that in the open sea.

According to Bogorov (1969), the total productivity in the ocean amounts to approximately 55×10^{15} g yr^{-1}, corresponding to 0.015 g cm^{-2} yr^{-1}. Thus, productivity in the ocean as a whole, and particularly per unit of its surface, is appreciably less than that on land.

Table 2.2
Productivity of Natural Vegetation Cover

Continent	Productivity (dry organic matter)	
	$(\times 10^{-15})$g	g/cm^{-2}
Europe	8.9	0.085
Asia	38.3	0.098
Africa	31.0	0.103
North America	18.1	0.082
South America	37.2	0.209
Australia (including Oceania)	7.6	0.086
All land	141.1	0.095

From these data it follows that the annual productivity for the entire globe is about 200×10^{15} g yr^{-1} or 0.04 g cm^{-2} yr^{-1}. At such a level of productivity, organic matter uses about 300×10^{15} g of CO_2 annually, i.e., over 10% of the atmospheric CO_2 content. Nearly all of this mass of CO_2 returns to the atmosphere and the hydrosphere by the oxidation of organisms and their waste products.

The rate of the photosynthetic cycle of atmospheric CO_2 depends largely on plant productivity on the continents. Considering this, one can find from the above data that the CO_2 cycle in the atmosphere is about 10 yr.

It is worth noting that a certain portion of organic carbon returns to the atmosphere as methane (CH_4) and carbon monoxide (CO) through the oxidation of organisms and their waste products. Methane is produced mainly in marsh and swamp areas, in flooded rice fields and, perhaps, in the ocean. Once in the atmosphere, CH_4 is oxidized and converted into CO. In addition to this source, a relatively small amount of CO is formed directly by decomposition of organic material and by fuel combustion. Carbon monoxide, in turn, is oxidized and transformed into CO_2. The available estimates have shown that the amount of CO_2 that is produced from CO is approximately 2×10^{15} g yr^{-1} (Garrels *et al.*, 1975).

Photosynthetic activity and oxidation of organic matter are particularly important in spatial temporal variations of CO_2.

Observations of the CO_2 distribution above a plant canopy have revealed a considerable CO_2 decrease in the lower air layer, especially near the top of the plant cover, if intensive photosynthesis occurs. Photosynthesis is also often responsible for a perceptible diurnal variation of the CO_2 concentration above the canopy, where the concentration reaches its minimum during the day, when photosynthesis is most intensive. While diurnal CO_2 fluctuations occur for the most part in the air layer near the earth's surface, annual fluctuations extend to a considerable height and spread over vast areas. Observations carried out at Mauna Loa (Hawaii) and at other stations have shown that the CO_2 concentration decreases by a few percent in the months with the most intensive photosynthesis as compared to its values in the seasons when photosynthesis is decreased.

The total mass of living organic matter that is maintained through the photosynthetic activity of green plants is only approximately known. Animal biomass is a small fraction of the total mass of living things, which consists mostly of plants, with trees a major portion. Because of this, the world biomass is determined largely by the expanse of forests on the continents.

According to estimates by Kovda (1969), the total biomass on the continents is roughly 3×10^{18} g. The carbon content in terrestrial plants as estimated by Kester and Pytkowicz (1977) amounts to 0.83×10^{18} g, corresponding to about 2×10^{18} g of dry organic matter. This estimate agrees reasonably well with that adopted by Kovda. If we consider that the land-plant biomass differs little from the global biomass, we find that the carbon cycle in living organisms lasts on the average 10–15 yr, approaching the mean cycle in the atmosphere.

A considerable amount of carbon is contained in products of organic decay on the continents, where the carbon mass in the detritus amounts to approximately 1.1×10^{18} g (Kester and Pytkowicz, 1977). We can see that the atmospheric carbon mass is comparable to that in living organisms and the products of their decomposition on the continents.

It is important to emphasize that the carbon cycle that results from the creation of organic material by plants and its oxidation is almost completely closed. Of the total amount of organic carbon assimilated annually by the plant, only a very small portion goes into the lithosphere and leaves this cycle.

Carbon exchange between the lithosphere and the atmosphere

In the sedimentary layers of the earth's crust there is a considerable amount of carbon, corresponding to 245×10^{21} g of CO_2. Organic carbon represents about one fifth of this mass, amounting to $\sim 12.5 \times 10^{21}$ g, of which 9.2×10^{21} g is contained in the continental crust (Ronov and Yaroshevsky, 1976).

There are two hypotheses relating to the formation of carbon compounds in sediments. The first is concerned with gradual release from the atmosphere, which contained a large quantity of CO_2. The second suggests that the source of carbon should be CO_2 entering the atmosphere by a continuous outgassing of the upper mantle as a result of its heating during the decay of long-lived radioactive isotopes.

Enough proof exists to adopt the second of these hypotheses. For example, Ronov has emphasized that if the ancient atmosphere contained enough carbon to form the sediments, the CO_2 pressure in the atmosphere would have been very high. Under these conditions the natural environment on the earth's surface would have been similar to that of Venus and this is at variance with the available data on the geological past of the earth (Ronov, 1976).

Outgassing of the upper mantle is probably associated with thermal convection. As a result, the degassing rate varied considerably in the past. Gas ejection from the depths of the lithosphere by volcanic eruptions seems to be the most noticeable emission of gases into the atmosphere. In addition to volcanic ejections, many gases undoubtedly enter the atmosphere by other means such as hot springs.

Free CO_2 from the atmosphere and the hydrosphere returns to the lithosphere in different ways. The greatest mass of carbon enters the lithosphere as a result of organic carbon deposition and formation of carbon rocks.

In order to estimate the rate at which CO_2 enters the lithosphere we can use Ronov's geochemical data (1976). Since little data exist on sediment formation in the Quaternary period, a CO_2 assimilation rate that corresponds closely to contemporary conditions can be obtained by using data on the last part of the Tertiary, the Pliocene.

We can conclude from Ronov's data that during the Pliocene the mean rate of organic carbon deposition was 0.014×10^{21} g per million years. This corresponds to about 0.05×10^{21} g $(10^6 \text{ yr})^{-1}$ of CO_2 consumption. Considering this estimate and using the above data on photosynthetic productivity, we can evaluate how much of the organic matter produced by green plants enters the lithosphere. This portion appears to be very small, only $\sim 0.02\%$.

From the data on CO_2 absorption rate we can learn how long it takes atmospheric CO_2 to be used up by the formation of sediments. This time appears to be approximately 30,000 yr.

In addition to the atmospheric CO_2 fixed in long-lived sedimentary rocks, a large amount of CO_2 is absorbed on the continents by reactions with rocks. Garrels and Mackenzie (1971) believe that this amount is 0.7×10^{15} g yr^{-1} and they emphasize the point that in the long run, this CO_2 will return to the atmospheric system as a result of reactions occurring in the ocean or in the deposited sediments. According to their evaluation, the atmospheric carbon cycle connected with the indicated process lasts about 3000 yr.

The above-mentioned estimate of organic carbon fixing can be used to compute the total rate of free O_2 release resulting from the activity of green plants. This release appears to be $\sim 0.04 \times 10^{15}$ g yr^{-1}. The O_2 that enters the atmosphere is consumed by various atmospheric and lithospheric processes such as weathering of minerals and oxidation of gases ejected by volcanic eruptions.

From these concepts concerning the gas exchange between the atmosphere and the lithosphere, it is clear that the total rate of atmospheric CO_2 consumption, which depends on its concentration, is not directly determined by the rate of CO_2 release into the atmosphere. In this connection the CO_2 cycle in the atmosphere–lithosphere system is not closed and consequently the amount of CO_2 in the atmosphere can increase or decrease over a given period of time. We can draw a similar conclusion concerning atmospheric O_2. This question is considered in more detail in the next subsection.

Atmospheric CO_2 thus takes part in two cycles, the first associated with the activity of the autotrophic plants and the second dependent on the gas exchange between the lithosphere and the atmosphere.

Although the first of these cycles has a far faster rate than the second, it affects the atmospheric CO_2 balance in a minor way because of the comparatively rapid oxidation of nearly all of the organic material produced by green plants. Despite a short period for replenishment of the CO_2 in the atmosphere by means of photosynthesis (of the order of 10 yr), the amount of atmospheric CO_2 changes relatively slowly. The rate of CO_2 consumption by organic carbon fixation in sedimentary rocks corresponds to a far longer period in the atmospheric CO_2 cycle—tens of thousands of years. Since the carbon cycle is not closed in the lithosphere–atmosphere system, gas exchange between the lithosphere and the atmosphere can cause large variations in the CO_2 mass over intervals that appear to be long in terms of human activity but are very short in the geological history of the earth.

2.3 VARIATIONS IN THE CHEMICAL COMPOSITION OF THE ATMOSPHERE

Geological time scale

In considering the question of variations in the chemical composition of the atmosphere in the geological past, it is necessary to use data on the earth's history that are related to the geochronological scale presented in Table 2.3. This scale is based on investigations of the earth's crust revealing sequential variations in the nature of the lithosphere, flora, and fauna. To use this table some knowledge of the relationships of various periods of the earth's history to variations in living nature is essential.

Data on fossil organisms become scarcer the more remote the period. Although the Archeozoic and the Proterozoic were much longer than the Phanerozoic (which includes the Paleozoic, Mesozoic, and Cenozoic eras), the data available on Precambrian organisms is scanty compared to that on Phanerozoic organisms.

The photoautotrophic plants are supposed to have evolved about 3 billion years ago. Traces of Precambrian plants and animals are extremely scarce because of the absence of hard tissues or possibly because of the almost complete disintegration of the remains of such ancient organisms in the process of rock metamorphism.

Table 2.3
Geochronologic Time Scale

Era	Period	Epoch		Duration (millions of years)	Absolute age (millions of years)
Cenozoic	Quaternary	Holocene Pleistocene		2	2
	Tertiary	Pliocene Miocene	Neogene		
		Oligocene Eocene Paleocene	Palaeogene	64	66
Mesozoic	Cretaceous			66	132
	Jurassic			53	185
	Triassic			50	235
Paleozoic	Permian			45	280
	Carboniferous			65	345
	Devonian			55	400
	Silurian			35	435
	Ordovician			55	490
	Cambrian			80	570
Proterozoic Archeozoic					

Data on the existence of organisms in the Archeozoic are mostly indirect. For example, the Archean carbonaceous sediments could have resulted from the activity of bacteria and algae absorbing mineral matter from ocean waters. Discoveries have been made of traces of Proterozoic marine organisms such as blue-green algae or diverse species of animals including upper Proterozoic multicellular organisms—coelenterata.

Numerous water-dwelling animals existed in the Cambrian period, comprising a great many species of trilobites, brachiopoda, archaeocyatha, crustacea, various molluscs, medusas, and others. The Cambrian vegetation included many kinds of algae.

More species of invertebrates appeared in the Ordovician and Silurian periods. In addition to trilobites, numerous arthropods evolved, including ancient crabs and scorpions. Some of the arthropods were the first inhabitants of the land. At the same time the first vertebrates—primitive agnathous fish—evolved. In the Ordovician and Silurian land vegetation developed, including various Psilophytineae.

Great changes in flora and fauna were characteristic of the Devonian period. At that time Crossopterygians emerged, then Amphibians originated from them. At the end of the Devonian the first insects developed. In addition to new groups of animals, new plants appeared. Instead of the ancient Psilophytineae, the Equisetaceae, Felicales, and Gymnospermae evolved. These flora grew vigorously in the Carboniferous period when forests spread widely; later their remains formed the coal deposits. A great variety of land insects developed during this period, some of which were gigantic. In the late Carboniferous the first reptiles appeared.

Gymnospermae (cycadicae and gingkoinae) are characteristic of the Permian and were widely distributed at the end of this period. At the same time amphibians thrived and reptiles advanced rapidly.

At the threshold of the Mesozoic era the organic world changed considerably. Many species of marine invertebrates, trilobites included, became extinct and others grew rare.

The Triassic marked a period of exuberant growth of reptiles, which extended throughout the entire Mesozoic era. Tortoises, dinosaurs, and marine reptiles such as ichthyosauri and plesiosauri appeared in the Triassic. The first primitive mammals evolved toward the culmination of that period.

In the Jurassic period gigantic amphibians became extinct and a large number of giant reptiles emerged, including various dinosaurs. During this period flying reptiles, pterodactyls, flourished. At the same time the first birds (archeopteryx), which strongly resembled the reptiles, also evolved. During this period and in the subsequent Cretaceous period there were some mammals but they were not notable representatives of the then existing fauna.

In the Cretaceous period snakes appeared as well as new species of dinosaurs and large flying reptiles.

During most of the Mesozoic era coniferous plants and ferns prevailed on the continents. In the mid-Cretaceous the first angiospermous plants sprang up. These spread rapidly over the continents forming forests of oak, poplar, beech, palm, etc.

The beginning of the Cenozoic era is a turning point in the history of the organic world. The overwhelming majority of reptiles became extinct in the late Cretaceous, including land dinosaurs and marine reptiles (ichthyosauri and plesiosauri) as well as flying reptiles. Of the numerous Mesozoic reptiles, only crocodiles, turtles, lizards, and snakes survived as part of the fauna of the Cenozoic.

Placentalia, which appeared in the Cretaceous, advanced rapidly in the early Tertiary period. Soon mammals occupied the ecological niches vacated by the vanished reptiles. Many mammals grew to enormous proportions. Some groups settled in the ocean while others acquired the ability to fly. In addition to mammals, birds were widely distributed during the

Cenozoic. They also occupied some of the ecological niches inhabited earlier by reptiles.

Throughout the entire Cenozoic era relatively small changes occurred in vegetation. With climatic changes, however, zones of different kinds of vegetation moved over long distances, particularly in the Quaternary period.

The late Cenozoic was marked by the appearance of *Homo sapiens*. Although primates related to man had existed even in the Paleocene, they progressed relatively slowly for a long time. In the lower Oligocene the first anthropoid apes developed and in the Pliocene there appeared *Australopithecus*, akin to man, followed by primitive human beings. Modern man appeared 35 to 40 thousand years ago in the epoch of the last glaciation; his activity was vital in the evolution of the biosphere.

Oxygen and carbon dioxide balances

As mentioned above, O_2 and CO_2 are involved in the biotic cycle, which includes the interaction of these gases with living matter and its waste products. Oxygen in the biological cycle is produced by photosynthesis of autotrophic plants and consumed in the oxidation of organic material. According to the data given above, the annual production of dry organic matter via photosynthesis appears to be $\sim 2 \times 10^{17}$ g, of which about $\frac{3}{4}$ is produced by continental vegetation cover and about $\frac{1}{4}$ is produced in the ocean (Efimova, 1977; Bogorov, 1969). This process produces 2×10^{17} g of O_2 annually.

Nearly the entire mass of O_2 originating from green plants is assimilated by the mineralization of organic material. The difference between O_2 production and consumption in the biotic cycle corresponds to the amount of organic carbon which accumulates in sedimentary rocks.

The relatively small differences between production and consumption of O_2 in the biotic cycle were very important in the evolution of the atmosphere. Variations in the atmospheric O_2 mass over a given period correspond to the difference between production (which is determined by the total input and output in the biotic cycle as well as by photodissociation of water vapor in the atmosphere) and consumption (mainly by oxidation of various minerals and gases from the earth's crust).

Undoubtedly one of the most important features of the atmospheric O_2 balance in Phanerozoic times was the dominance of O_2 production over consumption as a result of autotrophic plant activity. The problem of how O_2 from the dissociation of water vapor by ultraviolet radiation affects this balance is far more complicated.

Opinions on quantitative estimates of photodissociation have basic differences. For instance, Berkner and Marshall (1965a, 1966) contend that, without photosynthesis, the amount of O_2 produced by water vapor photodissociation could not exceed 0.1% of the modern value. To obtain this estimate it was taken into consideration that a very small quantity of O_2 will absorb the most intense portion of ultraviolet radiation. As a result, a screening effect of O_2 makes it impossible for a considerable mass of it to accumulate.

There are also a number of sources (Byutner, 1961; Brinkman, 1969) with higher estimates for the water vapor photodissociation effect, comparable to the results for photosynthesis.

Table 2.4
Continental Sediments

Geological period	Absolute age (millions of years)	Mass (g/million yr) $\times$ 10^{-21}		
		Igneous rocks	CO_2 total in sedimentary rocks	Organic carbon in sediments
Lower Cambrian		0.36	0.45	0.009
Middle Cambrian	490–570	0.18	0.32	0.005
Upper Cambrian		0.14	0.32	0.003
Ordovician	435–490	0.33	0.27	0.009
Silurian	400–435	0.19	0.22	0.005
Lower Devonian		0.62	0.24	0.003
Middle Devonian	345–400	1.43	0.66	0.018
Upper Devonian		1.76	0.68	0.024
Lower Carboniferous		1.02	0.70	0.022
Middle and Upper Carboniferous	280–345	0.27	0.30	0.010
Lower Permian	235–280	0.95	0.61	0.009
Upper Permian		0.35	0.23	0.005
Lower Triassic		0.32	0.18	0.004
Middle Triassic	185–235	1.10	0.34	0.004
Upper Triassic		0.85	0.28	0.009
Lower Jurassic		0.33	0.31	0.016
Middle Jurassic	132–185	0.38	0.39	0.033
Upper Jurassic		0.31	0.40	0.028
Lower Cretaceous	66–132	0.67	0.32	0.023
Upper Cretaceous		0.65	0.44	0.018
Paleocene		0.31	0.20	0.012
Eocene		0.37	0.31	0.020
Oligocene	2–66	0.17	0.08	0.017
Miocene		0.35	0.18	0.019
Pliocene		0.32	0.09	0.014

Most authors of contemporary surveys (Rutten, 1971; Garrels and Mackenzie, 1971; Cloud, 1974) are convinced that only very small concentrations of atmospheric O_2 can be maintained by photodissociation, since biological and geological processes in the epoch preceding the appearance of autotrophic plants reveal an almost complete absence of O_2 in the ancient atmosphere. From this standpoint, photosynthesis seems to be the basic source of atmospheric O_2 during the Phanerozoic.

Information on O_2 production in the course of the biotic cycle during different periods can be obtained from the data concerning organic carbon in sedimentary rocks.

Carbon dioxide generally enters the atmosphere and the hydrosphere from the earth's interior through volcanic eruptions. In the course of this process both CO_2 and CO are ejected, the latter being converted into CO_2 by interaction with atmospheric O_2. In addition to its being fixed in organic carbon sediments, CO_2 is vitally important in carbonate formations and the weathering of different minerals.

Our estimation of free atmospheric and hydrospheric CO_2 consumption during various periods can be based on the total CO_2 mass in sedimentary rocks of different ages.

The data necessary for studying atmospheric evolution in Phanerozoic time have been obtained in Ronov's investigations of sedimentary and igneous rocks on the continents (Ronov, 1959, 1964, 1976). These data are shown in Table 2.4. It has been revised to some extent as compared to previous publications.

Composition of the atmosphere in Phanerozoic time

This section will give results obtained in the study of the evolution of the atmospheric chemical composition in the Phanerozoic that were carried out by the authors, Ronov and Yanshin (Budyko, 1977; Budyko and Ronov, 1979). According to Table 2.4, Phanerozoic sediments contained 7.3×10^{21} g of organic carbon, which corresponded to the formation of 19×10^{21} g of O_2.

Since the indicated data cover only the continental surface, the actual emission of O_2 during Phanerozoic time exceeded our estimates because of photosynthesis occurring in the ocean. The difference, however, is believed to be relatively small. As mentioned previously, the total photosynthesis in water reservoirs in the modern epoch is considerably less than that on the continents. The available data show that the organic carbon mass in oceanic sediments over the last 150 million years is far less than that on the continents (Ronov, 1976).

Since the highest rate of assimilation in the hydrosphere generally occurs in shallow reservoirs within the continental areas, the data on continental organic carbon sediments is believed to roughly reflect the dynamics of O_2 production on a global scale during the Phanerozoic.

Since the present atmosphere contains 1.2×10^{21} g of O_2 we recognize that more than 90% of the total O_2 production in the Phanerozoic was used in the oxidation of mineral compounds.

The variation of atmospheric oxygen with time, dM/dt is determined by the equation

$$dM/dt = A - B, \tag{2.1}$$

where A and B are O_2 production and consumption rates. The O_2 production rate A for different periods of time can be calculated from the data on organic carbon in sediments.

To estimate O_2 consumption B by oxidation of mineral matter, we should bear in mind that the rate was not constant but was dependent on the atmospheric O_2 mass, tending to zero when there was no O_2 in the atmosphere and increasing with the expansion of the O_2 mass. Taking this into account we can assume $B = \alpha M$, where α is a proportionality coefficient.

Assuming that $M = M_0$ at $t = 0$, we find from (2.1)

$$M - (A/\alpha) = (M_0 - (A/\alpha))e^{-\alpha t}. \tag{2.2}$$

We can estimate O_2 variations during the Phanerozoic by proceeding from Eq. (2.2) and the data on organic carbon presented in Table 2.4. In this estimation, by assuming a certain amount of O_2 in the early Cambrian, it is possible to calculate variations during the Lower Cambrian. In the subsequent calculation the value obtained for the end of the Lower Cambrian is assumed to be the initial O_2 amount in the Middle Cambrian, etc. Thus all necessary values of the parameter M_0 are obtained by the calculations, except the first one, for the early Phanerozoic.

It can be shown that the initial O_2 amount assumed has only a slight effect on the calculated results. For example, if we assume the O_2 amount at the beginning of the Cambrian to be either zero or the modern value, i.e., the two extreme assumptions, the results practically coincide for all the subsequent periods since the Ordovician.

Taking this into consideration, we assume that during the Lower Cambrian dM/dt was considerably less than the O_2 production in the biotic cycle. Therefore, throughout this interval, $M = A/\alpha$.

To calculate O_2 variations it is necessary to determine α. We suppose that this parameter varied with time, increasing in the epochs of active volcanism and orogenesis. Since it is difficult to take these variations into account, we confine ourselves to a mean value of α that can be obtained by different methods based on the available empirical data. For instance, α is 3.1×10^{-8} yr^{-1} if the O_2 mass obtained for the end of the Phanerozoic is equal to its present-day value, known from direct observations.

This method for determining the O_2 mass includes a hypothesis that the above data on organic carbon deposition approach their global values. To see whether the results are dependent on that kind of supposition, we can carry out a similar computation considering the data on sediments for each time interval to be proportional to a global value but representing only a small portion of it. In this case, the rate of O_2 variation will be low compared to its production or consumption rate and the O_2 mass M' appears to be proportional to the organic carbon mass for the appropriate period.

To define variations in atmospheric CO_2, we assume, by analogy with the computation carried out for O_2, that they depend on the total CO_2 being fixed in sedimentary rocks. We assume also that this value approaches zero when the atmospheric CO_2 concentration is negligible and increases with its rise. Hence it is natural to suppose that the carbon mass in sediments over a unit time interval is roughly proportional to the atmospheric CO_2 concentration.

In this case the proportionality factor can be found by comparison of the present-day CO_2 concentration and the amount of CO_2 in sediments in the late Quaternary period, the latter obtained by extrapolation of the variations during the Tertiary period from the data in Table 2.4.

Calculated variations in O_2 and CO_2 during the Phanerozoic are presented in Fig. 2.1. The curves M and M' correspond to O_2 fluctuations calculated by the previous two methods. We can see that the curves are rather similar, which shows that our assumption concerning the global character of the data used has an insignificant effect on the calculation results.

It is seen from the figure that the amount of atmospheric O_2 in the early Phanerozoic represents about $\frac{1}{3}$ of its present amount. Throughout the Phanerozoic the O_2 mass was increasing but the process was uneven.

The O_2 mass increased abruptly for the first time in the Devonian and Carboniferous periods, when it reached its present-day level. At the end of the Paleozoic the O_2 mass decreased and in the Triassic it approached the level of the first part of the Paleozoic. In the mid-Mesozoic a second sharp rise in O_2 concentration took place. Afterwards it gradually decreased.

It is worth noting that the increase in O_2 in the Devonian and the Carboniferous was accompanied by the expansion of continental vegetation cover. A considerable decrease in O_2 in the Permian and most of the

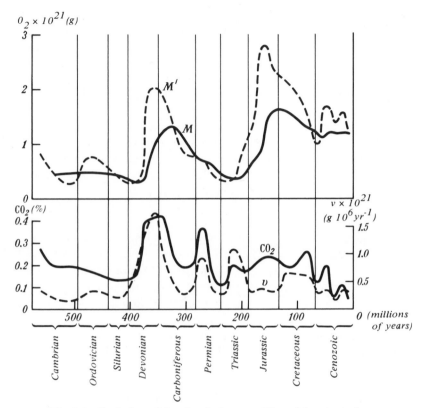

Fig. 2.1 Evolution of the chemical composition of the atmosphere.

Triassic is possibly linked to the arid conditions spreading over the continents at that time.

A high O_2 content in the atmosphere of the second half of the Mesozoic and the Cenozoic corresponded to an increase in autotrophic plant productivity due to the expansion of their evolving forms. A trend toward a slight decrease in O_2 since the Cretaceous period obviously reflected the effect of reduced CO_2 on vegetative productivity.

The curve corresponding to CO_2 fluctuations shows that its concentration varied within a range of 0.1 to 0.4% during most of the Phanerozoic. In the late Mesozoic the CO_2 concentration started to decrease gradually. The process was uneven, repeatedly giving rise to temporary increases. The last decrease in CO_2 concentration took place in the Pliocene. As a result, the present concentration has reached an unprecedented low level, which is several times lower than the Phanerozoic mean concentration.

To clarify the causes of fluctuations in atmospheric CO_2 content, we should study data on igneous rock formation per unit time interval. These data are presented in Fig. 2.1 in the form of the curve v. Since variations in igneous rock mass are considered to correspond to volcanic activity fluctuations, we conclude that great changes in volcanicity took place during Phanerozoic time. As shown in Fig. 2.1, volcanic activity varied rhythmically, its highest peaks occurring approximately every 100 million years.

According to Tikhonov *et al.* (1969), who studied the theory of convection in the lithosphere, this process is characterized by rhythmical upheavals of heated substances to the earth's surface at intervals of about 100 million years. It is noteworthy that this conclusion has been confirmed by the empirical data presented in Fig. 2.1.

Comparison of the curves v and CO_2 shows that fluctuations in volcanicity undoubtedly influenced the CO_2 content of the atmosphere. The highest peaks of volcanic activity were accompanied by increases in CO_2. It is interesting to note that comparatively small fluctuations in volcanicity during the Cenozoic also affected CO_2 content. A decreasing tendency of volcanic activity that prevailed since the mid-Cretaceous and corresponded to a decrease in CO_2 was of great importance in the evolution of the atmosphere in the late Phanerozoic.

In spite of the indicated common features of the variations in igneous rock formation and CO_2 content, there are certain differences between them. This is probably explained by some inaccuracy in the data used in the calculations but mainly by the fact that the mass of igneous rocks is only an indirect indicator of CO_2 entering the atmosphere from the earth's crust.

We can compare our data on the evolution of the composition of the atmosphere during the Phanerozoic with some concepts concerning the nature of this development, which have been presented in a few modern studies. It should be borne in mind that these suppositions are based on indirect considerations that have not been substantiated by quantitative calculations.

Surveys of atmospheric evolution are generally based on the previously cited works of Berkner and Marshall, who supposed that in the early Phanerozoic O_2 content was 1% of its contemporary value, in the Devonian period it rose to 10%, and in the Carboniferous it reached a maximum of 2–3 times its present-day value.

It is worth noting a hypothesis relating to O_2 variations in the post-Carboniferous period presented by Berkner and Marshall (1965b). A diagram in this paper shows that after the O_2 concentration reached its maximum in the Carboniferous, it decreased in the Permian to a small fraction of the modern concentration. Then the O_2 mass started increasing and in the Mesozoic it approached the present-day level.

If we compare this supposition with our results we shall see that in both cases the data agree reasonably well in a qualitative way. As for the quantitative estimates of O_2 fluctuations, those of Berkner and Marshall appear to be greatly overestimated. It should also be mentioned that the quantity of O_2 assumed in their work for the beginning of Phanerozoic time is strongly underestimated (by a factor of ~ 30). This error is easily explained since Berkner and Marshall made their estimates with the assumption that there were no multicellular organisms in the Precambrian period. From their point of view, this fact accounted for the prevalence in the atmosphere of anaerobic processes that proceed with a small quantity of O_2. As shown by present-day studies, a variety of multicellular organisms existed in the late Proterozoic. This invalidates the assumption of Berkner and Marshall.

In recent years some authors (Rutten, 1971; Cloud, 1974) using the data of Berkner and Marshall have not assumed that the early Phanerozoic was marked by such a small O_2 concentration, although their estimates of this concentration are also too low.

The article on the origin of the atmosphere by Cloud (1974) in the Encyclopaedia Brittannica has emphasized that great O_2 mass changes undoubtedly took place during Phanerozoic time but their nature is still unknown.

An opinion on atmospheric CO_2 fluctuations expressed long ago states that CO_2 concentration was gradually decreasing during Phanerozoic time. Some authors overestimated the range of CO_2 mass variations because they considered CO_2 one of the main air constituents in the early Phanerozoic. More correct estimates of CO_2 fluctuations were presented by Rutten (1971), who believed that the CO_2 concentration in the Phanerozoic was never more than 10 times the modern level. This supposition agrees well with the results of our study.

We should like to mention here Garrels' concept of the evolution of atmospheric CO_2, which is most fully set forth in his Russian paper (Garrels, 1975). In this work he emphasizes that in the Phanerozoic the lowest CO_2 concentration was at least $\frac{1}{3}$ of the contemporary value because autotrophic plants could not exist with lower levels of CO_2. Though the upper limit of CO_2 partial pressure is difficult to determine, it probably did not exceed 0.1 atm. In this connection Garrels refers to many other authors who believe that the upper limit of CO_2 concentration exceeded the present-day one by no more than a factor of 100.

Further, Garrels holds that the possible range of atmospheric CO_2

variation was not so wide. He thinks that the rate of CO_2 release into the atmosphere changed by no more than 9 times and that CO_2 fluctuations were dependent on variations in organic matter deposited as a result of photosynthetic fluctuations that corresponded to those of CO_2. Assuming that the photosynthetic rate was proportional to CO_2 concentration, and considering the lower limit of the latter to be 0.01%, Garrels comes to the conclusion that during the Phanerozoic the atmospheric CO_2 content varied within a range of 0.01 to 0.09%.

This conclusion is similar to the above statement that CO_2 concentration changed by approximately 10 times at this period. However, according to Garrels, the lower and upper limits of these variations are about $\frac{1}{3}$ of our estimates.

This difference can be easily explained since Garrels' estimates of the lower CO_2 limit correspond to conditions of complete or almost complete interruption of autotrophic plant activity although no data imply that such a situation occurred in the past. Moreover, such a concentration could not have existed during the last few hundred million years, because if it had, a drastic ecological disaster would have taken place, traces of which would have remained in the geological annals.

Results of our calculations of CO_2 concentration in the past proved to be in good agreement with Garrels' inferences if the value of the lower level of CO_2 concentration is assumed to be, not 0.01%, but 0.03%, i.e., the contemporary value.

As to the relationships between atmospheric CO_2 and O_2 evolution, an opinion was expressed in a number of papers that atmospheric CO_2 decreases with an increase in O_2 and increases with its decrease in the atmosphere (Ronov, 1959; Sochava and Glikman, 1973). The results obtained here show that in some cases this supposition does not hold true.

Quantitative calculations of the chemical composition of the atmosphere for the entire history of the earth have been made recently by Hart (1978), whose work is discussed in more detail in Chapters 4 and 6. Hart has not used empirical data on the composition of sediments in his calculations of O_2 and CO_2 variations. As a result, some of his conclusions appear to be unrealistic. For instance, assuming that in the Phanerozoic carbonates were not produced at the expense of the degassing of the lithosphere, Hart arrives at the conclusion that the atmospheric CO_2 concentration did not vary at that time. Since most of the carbonates were actually formed from gases ejected by the lithosphere in the Phanerozoic, this conclusion can not be considered correct.

In summarizing the discussion of variations in the chemical composition of the atmosphere during the Phanerozoic, we should mention certain factors that limit the accuracy of the above data on O_2 and CO_2 concentrations in the ancient atmosphere. For instance, the available data on the mass of carbon in sediments are incomplete. However, we believe that in the future this kind of computation will be based on more complete information, covering not only the continents but the oceans as well.

The estimation of CO_2 content in the geological past is very important in the study of climatic changes. We think that the above data on CO_2 variations as a rough approximation, reflect the actual evolution of the composition of the atmosphere. This supposition is based on a few independent grounds. In the first place, the highest CO_2 concentrations obtained by our calculations occurred mostly during the epochs of highest volcanicity. Secondly, the results given below on climatic change caused by CO_2 fluctuations agree reasonably well with the empirical data on climatic conditions in the past that are presented in Chapter 4. Thirdly, the mean CO_2 concentrations obtained for the Phanerozoic seem to approach the optimum concentration for photosynthesis. This confirms the idea that autotrophic plants adapted to the chemical composition of an atmosphere that was richer in CO_2 than the present one. Fourthly, a range of CO_2 variations obtained by calculation agrees well with earlier suppositions, from the regularities of sediment formation, as to how that concentration could vary.

Effects of variations in atmospheric composition on living nature

From the aforementioned data we can see that the Phanerozoic CO_2 concentration in the atmosphere varied by a factor of ~ 10 and that of O_2 by a factor of 4.

Considering the results of physiological investigations, we believe that even if CO_2 fluctuations did not exert a permanent effect on the vital activity of animals, they were of great importance for the photosynthetic processes of autotrophic plants.

It has been established in laboratories and field investigations (Rabinovich, 1951) that the modern CO_2 content of the atmosphere is only about one-tenth of that necessary to reach the highest productivity of photosynthesis in many autotrophic plants. In this connection, as mentioned above, there was proposed long ago a theory that the basic groups of autotrophic plants appeared in the epoch with a higher concentration of CO_2 in the atmosphere. This supposition was entirely confirmed by the data of this study, leading to the conclusion that for almost the entire Phanerozoic the concentration of atmospheric CO_2 was considerably higher than it is today. On the other hand, it never reached the level of a few percent, when the productivity of photosynthesis of many plants starts to decrease. Apparently a reduction in the quantity of CO_2 in the late Tertiary vitally influenced living nature. This question is treated later.

We believe that fluctuations in O_2, unlike those of CO_2, affect animals more than plants. We should mention here that the role of O_2 variations in the evolution of animals during the Phanerozoic has been discussed in a number of papers (for example, MacAlester, 1970; Schopf *et al.*, 1971; Sochava and Glikman, 1973). It was difficult to study this problem in the past, for relatively reliable data on variations in O_2 during Phanerozoic time were not available.

It should be kept in mind that an increase in O_2 was favorable for animals of the following groups, other things being equal.

(1)   Animals that expend much energy on movement. This includes land animals compared with water-dwelling ones (for which the influence of gravity is relatively small) and flying animals compared with land and water-dwelling species. Large animals expend comparatively more energy on movement than small ones, other things being equal. Considering land animals, we can say that more energy is expended by animals with a higher position of the center of gravity (for example, bipeds compared with quadrupeds).

(2)   Large, and especially gigantic, animals. The level of metabolism per unit weight of animals having other features in common diminishes, as a rule, with increase in their size, which results in difficulty in providing the internal organs of large animals with O_2.

(3)   Animals with heat regulation compared with poikilothermal forms, because the former have a higher level of metabolism.

We should mention that the emergence of vertebrates on the land in the Devonian and the broad expansion of many species of amphibians and reptiles in the Carboniferous period took place in the epoch of sharp increase in O_2. In the Carboniferous, flying insects were broadly distributed; it is noteworthy that there were some gigantic forms among them.

The decrease in the amount of O_2 that was especially noticeable in the Triassic period was accompanied by the extinction of many species of Paleozoic land reptiles.

Animals with heat regulation developed after a significant rise in O_2 level: mammals in the late Triassic period, birds in the middle of the Jurassic. Although birds are morphologically closer to reptiles than are mammals, they appeared later. This is probably explained by the higher level of metabolism necessary for the great expenditure of energy during flight. A rapid rise in O_2 level in the first half of the Jurassic period made it possible to reach this metabolic level. It is interesting that the other group of flying vertebrates, the pterodactyls, also appeared during the Jurassic, when there was a large quantity of O_2 in the atmosphere.

It should be emphasized that the largest land vertebrates were widely distributed in the Jurassic and the Cretaceous, i.e., during the periods of highest concentration of atmospheric O_2. For instance, gigantic biped dinosaurs that lived at this time consumed a great amount of O_2 while moving.

It thus appears that variations in atmospheric O_2 content definitely influenced the evolution of the animal world.

Variations in O_2 were of less importance for the evolution of plants than for animals. Nevertheless we can suppose that a rapid development of land vegetation in the Devonian and the Carboniferous periods depended partly on the atmospheric O_2 increase that was produced by the vegetation itself.

The dependence of the whole mass of living matter on CO_2 concentration is noteworthy. The major part of the global biomass is the substance of the autotrophic plants, the amount of which depends upon photosynthesis. Taking account of the dependence of the rate of photosynthesis on CO_2 concentration, we come to the conclusion that in the geological periods when the amount of CO_2 was greater than it is now, the mass of land vegetation could exceed its present magnitude. This mainly concerns vegetation in zones with sufficient humidity. Thus, fluctuations in the CO_2 concentration were evidently accompanied by variations in the mass of autotrophic plants and, consequently, by variations in the total mass of living matter on our planet.

Chapter

3

Semiempirical Theory of Climatic Change

3.1 ENERGY BALANCE OF THE EARTH

Energy balance of the earth's surface

Semiempirical theory of the atmospheric thermal regime is based on the results of studying the energy (heat) balance of the earth's surface and the atmosphere.

The earth's heat balance was first studied in the 19th century when actinometric instruments were devised and the amount of solar radiation reaching the top of the atmosphere was estimated as a function of latitude and season. Between 1915 and 1940 W. Schmidt, A. Ångström, F. Albrecht, and S. Savinov determined the heat balance components of the earth's surface for individual regions of the globe.

The author and his co-workers constructed a series of world maps of heat balance components of the earth's surface for each month and mean annual conditions, which was published as the "Atlas of the Heat Balance" (Budyko, 1955). These maps were further improved and together

50

with several maps of heat balance components of the earth–atmosphere system were later published as the "Atlas of the Heat Balance of the Earth" (Budyko, 1963).

Since 1963 a considerable amount of material has been obtained from a number of newly established actinometric stations on the continents. Of particular importance was information obtained from oceanic radiation observations. This made it possible to improve the estimation of the characteristics of the oceanic radiation regime. Because of progress in studying turbulent diffusion, it was possible to improve the methods for computing the heat expenditure for evaporation and turbulent heat exchange between the earth's surface and the atmosphere over the oceans.

New improved maps of the heat balance components (Budyko et al., 1978a) have been constructed using recently obtained data and modern computational methods.

In constructing these maps, the heat balance equation of the earth's surface is used in the form

$$R = LE + P + B, \tag{3.1}$$

where R is the radiation balance of the earth's surface, LE the expenditure of heat for evaporation or the heat income from condensation on the earth's surface (L is the latent heat of vaporization and E the evaporation or condensation rate), P the turbulent heat flux between the earth's surface and the atmosphere, and B the heat flux between the earth's surface and the lower layers of reservoirs or soil.

In (3.1) the radiation balance R is assumed to be equal to $Q(1 - \alpha) - I$ (where Q is the total shortwave radiation reaching the earth's surface, α the albedo of the earth's surface, and I the net longwave radiation), R being positive for the radiation income to the earth's surface. Other terms in the heat balance equation are positive when they correspond to heat expenditure.

World maps of the heat balance components of the earth's surface for mean annual conditions constructed in the aforementioned study are shown in Figs. 3.1–3.5. The quantities of heat balance components are expressed in these maps in kilocalories per square centimeter year and in kilocalories per square centimeter month. It should be remembered that 1 kcal cm^{-2} month$^{-1} = 15.9$ W m^{-2}.

As seen from Fig. 3.1, annual values of total radiation on the globe vary from less than 60 to more than 200 kcal cm^{-2} yr^{-1}. The largest values of total shortwave radiation over both land and oceans are observed in high pressure belts in the Northern and Southern Hemispheres. The total radiation decreases toward the poles. Some decrease in the total radiation is also observed at equatorial latitudes because of frequent overcast sky

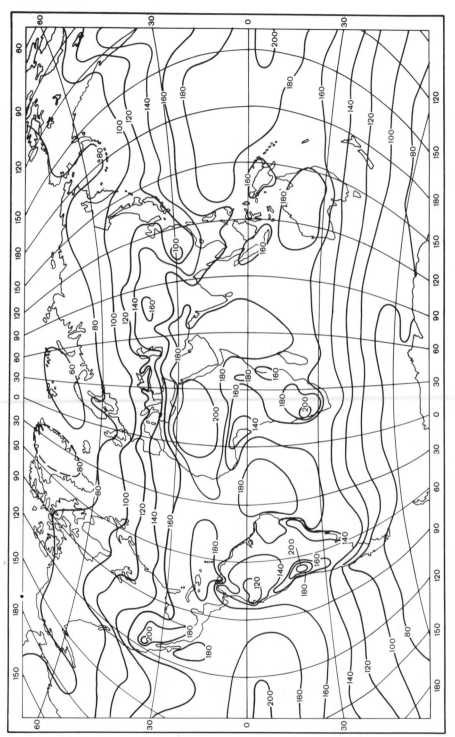

Fig. 3.1 Total radiation.

during the whole year. The distribution of total solar radiation isolines is mainly zonal. This zonality is considerably disturbed by uneven cloud distribution, especially in the mid-latitudes in both hemispheres where cyclonic activity is intensive (the west coast of Canada, northern Europe, the southwest coast of South America, etc.), in the eastern regions of tropical zones in the oceans under the influence of trade-wind inversions and cold marine currents, and in regions of monsoon circulation (Hindustan, the eastern coast of Asia, and northwest of the Indian Ocean).

Considering the data on total radiation distribution for the winter months, one should note that they are characterized by latitudinal variations in total radiation: from about 14 kcal cm^{-2} month^{-1} in low latitudes to zero in polar latitudes, where at this time there is no radiation. The largest monthly totals of solar radiation in low latitudes are observed in tropical monsoon regions where there is little cloudiness at this time of year.

The distribution of total shortwave radiation in summer is characterized by high values all over the hemisphere with little geographical variability. Tropical and subtropical deserts get maximum solar radiation, i.e., more than 20 kcal cm^{-2} month^{-1}. Large amounts of total shortwave radiation are observed in summer in the polar regions, where the greater length of the day compensates for small sun altitudes. The Antarctic central plateau gets the largest amounts of solar radiation in summer. For example, in January, monthly totals vary from 16 to 18 kcal cm^{-2} on the coast to 30 kcal cm^{-2} in the interior of the continent. These values noticeably exceed those for tropical deserts.

The radiation balance data for the earth's surface are presented in Fig. 3.2. The annual sums of the radiation balance at the earth's surface vary from values of less than -5 kcal cm^{-2} in the Antarctic and approximately zero in the central Arctic to 90 to 95 kcal cm^{-2} in tropical latitudes.

Due to the influence of astronomical factors, the annual and monthly radiation balance sums are zonally distributed over the flat territories in high and mid-latitudes in the Northern Hemisphere. The latitudinal distribution breaks down in regions where circulation factors significantly change cloudiness conditions as compared with the average.

Geographic zones in high and middle latitudes are characterized by definite annual radiation balance totals: Arctic deserts, less than 10 kcal cm^{-2} yr^{-1}; tundra, forest-tundra, 10–20 kcal cm^{-2} yr^{-1}; northern and middle taiga, 20–30 kcal cm^{-2} yr^{-1}; southern taiga, 30–35 kcal cm^{-2} yr^{-1}; mixed, leaf-bearing, and broadleaf forests, forest-steppe, and steppe in middle latitudes, 35–50 kcal cm^{-2} yr^{-1}.

In subtropical, tropical, and equatorial zones the regimes of the atmospheric circulation cause pronounced differences in moisture and cloudiness conditions resulting in annual radiation balance variations of from 55

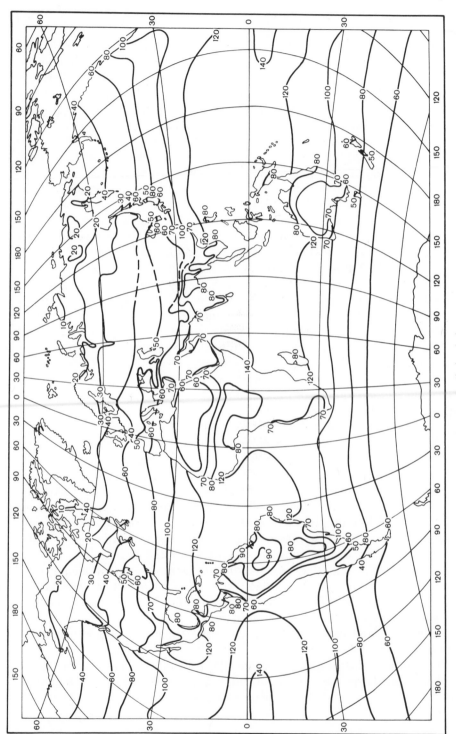

Fig. 3.2 Radiation balance.

to 95 kcal cm^{-2}. At the same time, the minimum values of the radiation balance in low latitudes are observed in subtropical and tropical deserts because of the high reflectivity of the deserts surface and considerable net longwave radiation heat losses under conditions of little cloudiness, low air humidity, and high soil surface temperature.

From radiation balance maps for individual months it can be seen that the lowest values of the radiation balance are observed in high polar latitudes—from -1 to -2 kcal cm^{-2} month^{-1} in winter and ~ 4 kcal cm^{-2} month^{-1} in summer. In mid-latitudes in the Northern Hemisphere the values of the radiation balance vary from -1 to -2 kcal cm^{-2} month^{-1} in January and from 7 to 9 kcal cm^{-2} month^{-1} in July. In tropical extraequatorial latitudes in the period of the winter solstice the values of the radiation balance are 3.5 to 4.0 kcal cm^{-2} month^{-1}; in summer the largest values reach 9 to 10 kcal cm^{-2} month^{-1}, decreasing to 5.5 to 6.0 kcal cm^{-2} month^{-1} in deserts and equatorial monsoon regions.

The distribution of radiation balance values on the surface of the oceans, as shown in Fig. 3.2, is similar to that of total radiation. The maximum value of the radiation balance on the oceans exceeds 140 kcal cm^{-2}yr^{-1}; the lowest values for ice-free ocean surface are at the boundary of floating ice, ~ 20–30 kcal cm^{-2} yr^{-1}. It should be emphasized that the annual totals of the radiation balance over the entire ice-free ocean surface are positive.

In winter the ocean radiation balance varies with latitude from 8 to 10 kcal cm^{-2} month^{-1} in low latitudes to small negative values of ~ -4 kcal cm^{-2} month^{-1} in high latitudes. In this case, the radiation balance becomes negative in both hemispheres, above 45°N in January and above 45°S in July.

In the summer months the values of the radiation balance of the oceans reach their maxima, exceeding 14 kcal cm^{-2} month^{-1} in low latitudes and 8 to 9 kcal cm^{-2} month^{-1} in high latitudes. In these months the radiation balance distribution differs noticeably from a zonal one, regions with high and low values corresponding to those with little and heavy cloudiness.

The annual map of latent-heat flux distribution for land and ocean is presented in Fig. 3.3. This map is constructed from calculations of long-term monthly means of evaporation (Drozdov *et al.*, 1974). Monthly means of the latent-heat flux (and turbulent heat exchange with the atmosphere) on the oceans are calculated using the data of marine climatic atlases and long-term ship observations in the Atlantic, Indian, and Pacific Oceans (Atlas, 1974).

Considering the peculiarities of the latent-heat flux distribution on land, it can be mentioned that its value varies by ~ 80 kcal cm^{-2} yr^{-1}. In regions of sufficient moisture, heat output for evaporation increases as the radia-

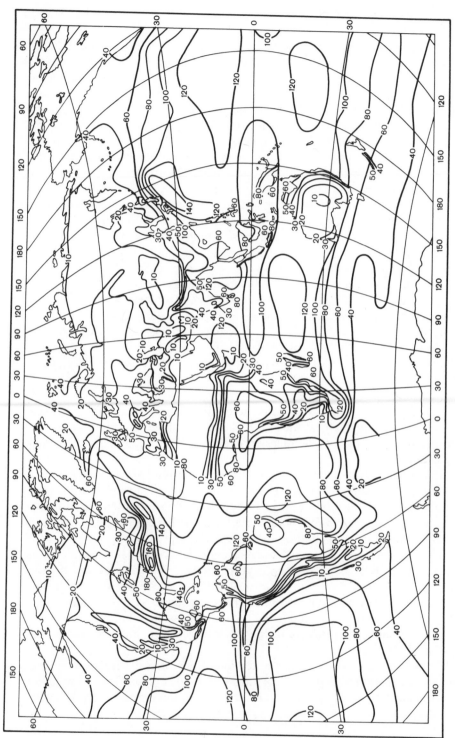

Fig. 3.3 Heat expenditure on evaporation.

tion balance increases toward the equator, varying from less than 10 kcal cm^{-2} yr^{-1} on the northern coasts of continents to 70 to 80 kcal cm^{-2} yr^{-1} in humid tropical forests in South America, Africa, and the Malay Archipelago. In regions with insufficient moisture, the value of heat outgo for evaporation depends on the aridity of the climate, decreasing with increasing aridity. The lowest values of the latent-heat flux are observed in tropical deserts where they amount to only a few kilocalories per square centimeter year.

The annual march of the heat outgo for evaporation is also determined by heat and moisture resources. In extratropical latitudes with sufficient moisture, maximum values of the latent-heat flux occur in summer, corresponding to the yearly variation of the radiation balance and reaching 5–6 kcal cm^{-2} month^{-1}. In winter the value of heat output for evaporation is small. In regions with insufficient moisture, the maximum heat output for evaporation is also usually observed during the warm period, but the time at which the maximum is reached depends strongly on the moisture regime. In deserts the annual march of LE is determined by that of precipitation.

In tropical latitudes with a humid climate the latent-heat flux is large throughout the year, amounting to 6–8 kcal cm^{-2} month^{-1}. In regions with seasonally reduced precipitation some decrease in heat output for evaporation is observed, but its variation in the course of the year is comparatively small. In regions with a clearly defined dry period, latent-heat flux maxima are observed at the end of humid periods and minima are observed at the end of dry ones.

For the land area as a whole (including Antarctica), the latent-heat flux is 27 kcal cm^{-2} yr^{-1}.

The distribution of the annual latent-heat flux on the oceans is, in general, similar to that of the radiation balance. As seen from the map (Fig. 3.3), the differences in latent-heat flux are rather large: the latent-heat flux varies from more than 120 kcal cm^{-2} yr^{-1} in tropical latitudes to $\sim$30 kcal cm^{-2} yr^{-1} at the edge of polar ice. In equatorial latitudes the flux is less than at higher latitudes (less than 100 kcal cm^{-2} yr^{-1}); this can be attributed to increased cloudiness and air humidity. In addition to the radiative heat expended in evaporation from the oceans, in a number of regions heat transferred by ocean currents also contributes to evaporation. Therefore, the zonal character of the latent-heat flux distribution is disturbed in some regions by the activity of warm and cold currents. This is clearly seen from the map in Fig. 3.3. The maximum latent-heat fluxes are observed in the Northern Hemisphere: >180 and $\sim$140 kcal cm^{-2} yr^{-1} in the Gulf Stream and Kuro Shio zones, respectively. In these regions, the evaporation is increased not only by high water temperature but also by the comparatively low humidity of the air masses incoming into these regions from North America and Asia during the cold season.

The annual total of heat outgo for evaporation from the oceans is mainly the sum of the values for the autumn–winter period.

The winter latent-heat flux distribution is similar to the annual distribution. In the cold season the effect of the warm currents is intensified. In this connection the distinctive features of each ocean show up most vividly: the expenditure of heat for evaporation from the surface of the North Atlantic in the mid-latitudes is twice as great as at the same latitudes in the Pacific Ocean. The lowest values of heat output for evaporation are observed in the mid-latitudes in the Southern Hemisphere in the Atlantic and Indian Oceans. Warmer air masses move from topical latitudes into the regions with comparatively low water temperatures, thus decreasing heat expenditure for evaporation.

In spring the effect of warm currents on the value of the latent-heat flux decreases due to a reduction in the energy resources of the currents. Since in summer the average wind speed and the water–air temperature difference both decrease, the latent heat outgo drops considerably. In addition, the difference in the values of heat outgo for evaporation from the surface of the individual oceans diminishes.

The annual turbulent heat flux map is presented in Fig. 3.4. Maximum sensible-heat outgo from the land surface to the atmosphere by the turbulent flux is noted in the tropical deserts, where it reaches $55–60$ kcal cm^{-2} yr^{-1}. The turbulent heat flux decreases with increasing moisture. In regions with humid tropical forests, the turbulent flux amounts to $10–20$ kcal cm^{-2} yr^{-1}. At higher latitudes the turbulent heat flux decreases as the radiation balance decreases. Over the northern coasts of the Northern Hemisphere continents the sensible-heat flux is less than 5 kcal cm^{-2} yr^{-1}. The same values are characteristic of regions in mid-latitudes that have sufficient moisture.

The same regularity is found in the annual variation: as the radiation balance increases, the turbulent flux also increases. Therefore, in extratropical latitudes, the largest values of sensible-heat flux are observed in summer, the smallest in winter. For areas north of 40°N and south of 40°S the direction of turbulent heat flux changes during the year. In winter the earth's surface gets heat from the atmosphere by means of turbulent exchange although the absolute values of the heat flux are small; even in high latitudes they do not exceed 1 kcal cm^{-2} month^{-1}.

The surface of the continents between 40°N and 40°S loses heat by the turbulent flux throughout the year. In low latitudes, the annual course of

Fig. 3.4 Sensible-heat flux from the earth's surface to the atmosphere.

the turbulent flux depends strongly on moisture. The largest monthly values of the turbulent flux are observed in the period of minimum precipitation. In subtropical latitudes with Mediterranean-type climate, the maximum values of turbulent flux occur in summer, reaching 6 kcal cm^{-2} month^{-1}. In deserts, particularly coastal deserts where the turbulent heat exchange is considerably influenced by the processes of air mass transformation at the water–land boundary, sensible-heat flux values exceed 6 kcal cm^{-2} month^{-1}. In humid tropical regions, the turbulent flux is small throughout the year, its monthly values being less than 2 kcal cm^{-2}.

Overall, 45% of the annual value of the radiation balance on the surface of the continents is expended in turbulent heat exchange with the atmosphere.

The map of turbulent heat exchange between the surface of the oceans and the atmosphere shows that almost everywhere the oceans lose heat. The greatest heat loss (as the heat outgo for evaporation) takes place in the western and the northwestern regions of the Northern-Hemisphere oceans, where the turbulent heat flux exceeds 40 kcal cm^{-2} yr^{-1}. At lower latitudes, particularly near the equator where temperature differences between the water surface and the air above it are small throughout the year, the turbulent heat flux from the surface of the oceans is less than 10 kcal cm^{-2} yr^{-1}.

In the Southern Hemisphere, because of the absence of sharp contrasts between the water and air temperatures, the turbulent flux is considerably less than that in the Northern Hemisphere and does not exceed 15 kcal cm^{-2} yr^{-1} anywhere.

Negative annual values of the turbulent heat exchange with the atmosphere (i.e., heat inflow from the atmosphere) are observed in the zones of the cold California Current and the Current of the West Winds in the Southern Hemisphere. These negative values have small absolute magnitudes.

The annual sums of turbulent heat exchange with the atmosphere, as well as those of heat outgo for evaporation, consist mainly of the values for the fall and winter periods. The closest relationship between heat exchange conditions and the effects of sea currents and atmospheric circulation is observed in winter.

Heat losses by the oceans in the winter months in the Northern Hemisphere reach values of 8–10 kcal cm^{-2} month^{-1}; in the Southern Hemisphere, about 5 kcal cm^{-2} month^{-1}.

In summer, for both hemispheres, the turbulent heat exchange between the surface of the oceans and the atmosphere is close to zero everywhere. Its values lie in the range of 1 to -1 kcal cm^{-2} month^{-1}.

The values of the heat exchange between the surface of the oceans and the deeper layers that are due to the effects of currents are the algebraic sums of the radiation balance, heat outgo for evaporation, and turbulent heat exchange between the surface of the oceans and the atmosphere. In this case, the algebraic sum of the terms entering the heat balance equation for the oceans as a whole amounts to 2% of the radiation balance value. This verifies that the methods used for calculating the components of the oceanic heat balance are sufficiently reliable.

Figure 3.5 shows the annual map of a residual component in the heat balance equation—the heat incoming to or outgoing from the ocean surface by the activity of marine currents. This map shows the role of currents and atmospheric circulation in heat exchange between the surface and the deeper layers of the oceans. In addition, this map characterizes heat redistribution between different ocean regions.

It can be seen that in equatorial and to some extent in tropical latitudes of both hemispheres, as a result of large radiative heat inflow and lowered outgo by evaporation and turbulent heat exchange with the atmosphere, ocean waters gain $20–40$ kcal cm^{-2} yr^{-1}. The higher the latitude in either hemisphere, the less the accumulation of heat by ocean waters. In high latitudes the amount of radiative heat proves to be insufficient to compensate for heat outgo by evaporation and turbulent heat flux, so heat transferred by ocean currents is directed from deeper layers toward the ocean surface. The areas with maximum heat outgo from the ocean surface into the atmosphere are located in the regions influenced by warm currents— the Gulf Stream and Kuro Shio. In these regions during the cold season the warm ocean waters interact with the fluxes of cold continental air from North America and Asia. This causes an intensive heat transfer from the deeper ocean layers, through the water surface, into the atmosphere. This ocean heat outgo amounts to -110 kcal cm^{-2} yr^{-1}.

It should be mentioned that in high and middle latitudes of the Northern Hemisphere, because of large water–air temperature contrasts, a similar process takes place in regions influenced not only by warm, but even by some cold, currents.

In the Southern Hemisphere, where the contrasts between the thermal state of the water and the air are considerably smaller, the heat loss from the ocean surface does not exceed in absolute magnitude -30 to -35 kcal cm^{-2} yr^{-1}.

In most regions of cold currents, the ocean surface gains heat from the sun and the atmosphere and transfers a significant portion to underlying layers. The maximum heat flux reaching deeper layers of the ocean, with a value of about 60 kcal cm^{-2} yr^{-1}, is observed in the region of the California

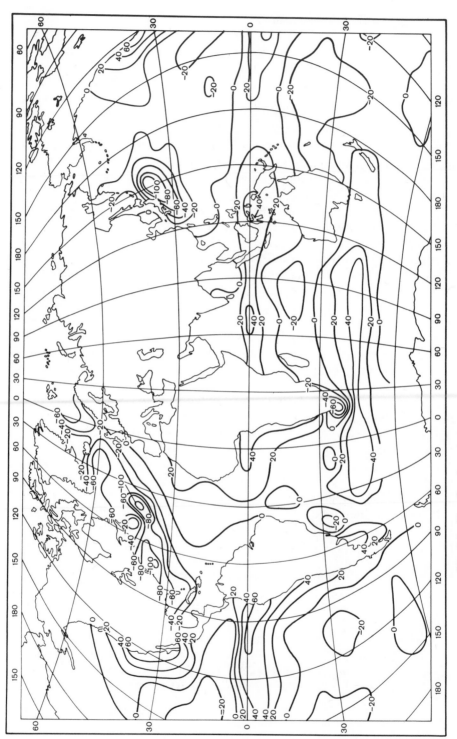

Fig. 3.5 Heat flux from the surface of the oceans to deeper water layers.

current. In the Southern Hemisphere the surface waters of the Peru, Benguela, Falkland, and West Winds currents transfer somewhat lesser amounts of heat to deeper layers—not more than 50 kcal cm^{-2} yr^{-1}.

The question of the accuracy of determining the heat balance components of the earth's surface using calculated results based on the data of ground meteorological observations is discussed, for example, in Budyko (1971). It is worth noting that Holloway and Manabe (1971) constructed world maps of the heat balance components of the earth's surface using the general circulation model of the atmosphere. In this investigation values of the radiation balance of the earth's surface, evaporation (which is proportional to heat outgo for evaporation), and the turbulent heat flux between the earth's surface and the atmosphere are determined. When these maps were compared with similar ones from "Atlas of the Heat Balance of the Earth" (Budyko, 1963), Holloway and Manabe found good agreement between them.

Oort and Vonder Haar (1976) used another method of determining some heat balance components of the earth's surface. In their study, they used the data of satellite observations of the radiation balance components of the earth–atmosphere system. Developing the idea used in their previous investigation, Oort and Vonder Haar computed the heat flux between the earth's surface and the atmosphere from the atmospheric heat balance equation. They considered the value of this flux to be equal to the algebraic sum of the radiation balance of the earth–atmosphere system, the change in the heat content in a vertical column of the atmosphere, and the horizontal heat inflow into this column caused by atmospheric circulation. The first of these three components was obtained from satellite data, the second and third from aerological observations. Using this method they computed latitudinal monthly means of the heat flux equal to the sum of surface radiation balance, heat outgo for evaporation, and turbulent heat flux between the earth's surface and the atmosphere. Taking into account the limited accuracy of determining from aerological observation data the heat inflow into the air column under the influence of atmospheric circulation and the fact that the residual term of the heat balance equation computed by Oort and Vonder Haar is often less in absolute magnitude than its principal components, one should note that the results obtained are fairly accurate. They found, in particular, that the value of the total heat inflow at different latitudes in the Northern Hemisphere varies from 40 to − 140 W m^{-2} in January and from 30 to 90 W m^{-2} in July. The same values obtained from the data of the "Atlas of the Heat Balance of the Earth" vary from 40 to − 110 W m^{-2} in January and from 10 to 90 W m^{-2} in July. This agreement shows that even approximate calculations of the heat balance components provide satisfactory results in spite of the fact that the determination of the above values as a residual term of the heat balance from the "Atlas of the Heat Balance of the Earth" gives a considerable error.

Values of the heat balance components of the earth's surface averaged over the earth as a whole can be obtained from data on their spatial distribution.

Table 3.1 gives the values of heat balance components for different latitudinal zones of land and ocean and for the earth as a whole. Data presented in this table can be compared with the results of previous studies of the heat balance of the earth's surface. Table 3.2 presents relative values of heat balance components (expressed as percent of the solar constant value) obtained from different studies.

From the data given in the last three lines of Table 3.2, it is seen that as the world maps of heat balance components improve, the values of absorbed radiation, radiation balance, and heat outgo for evaporation appear to increase somewhat. Relative values of heat balance components calculated in the last study (Budyko *et al.*, 1978) approach the results obtained by Houghton in the 1950s (except for the value of the turbulent heat flux).

Table 3.1

Latitudinal Means of Components of the Heat Balance at the Surface of the Earth[a]

Latitude (degrees)	Land			Ocean				Earth as a whole			
	R	LE	P	R	LE	P	F_o	R	LE	P	F_o
North											
70–60	22	16	6	23	31	22	−30	22	20	11	−9
60–50	32	23	9	43	47	19	−23	37	33	13	−9
50–40	45	25	20	64	67	16	−19	54	45	18	−9
40–30	58	23	35	90	96	14	−20	76	65	23	−12
30–20	64	19	45	111	109	7	−5	94	75	21	−2
20–10	74	32	42	121	117	7	−3	109	95	16	−2
10–0	79	57	22	124	104	7	13	114	93	10	11
South											
0–10	79	61	18	127	99	6	22	116	90	9	17
10–20	75	45	30	122	113	9	0	112	98	14	0
20–30	71	28	43	109	106	11	−8	100	88	18	−6
30–40	62	29	33	92	82	11	−1	88	76	14	−2
40–50	44	22	22	72	51	6	15	71	50	7	14
50–60	35	22	13	46	35	9	2	46	35	9	2
Earth as a whole	50	27	23	91	82	9	0	79	66	13	0

[a] Values are given in kilocalories per square centimeter year.

Table 3.2
Heat Balance at the Surface of the Earth[a]

Investigation	Absorbed shortwave radiation	Net longwave radiation	Radiation balance	Heat outgo for evaporation	Turbulent heat flux
Dines (1917)	42	14	28	21	7
Alt (1929)	43	27	16	16	0
Bauer and Phillips (1934)	43	24	19	23	−4
Houghton (1954)	47	14	33	23	10
Lettau (1954)	51	27	24	20	4
Budyko (1955)	42	16	26	21	5
Budyko (1963)	43	15	28	23	5
Budyko et al. (1978a)	46	15	31	26	5

[a] Components of the heat balance are expressed as percent of the amount of solar radiation incident on the outer boundary of the atmosphere.

The energy balance of the earth–atmosphere system

Components of the heat balance of the earth–atmosphere system were first calculated in the 1920s and 1930s by G. Simpson, F. Baur, H. Phillips, and G. Trolle. World maps of the earth–atmosphere system radiation balance and net heat income from condensation and advective processes in the atmosphere, constructed by Vinnikov, were published in "Atlas of the Heat Balance of the Earth" (Budyko, 1963). In the late 1960s, as a result of progress in satellite meteorology, maps of the radiation regime elements of the earth–atmosphere system were constructed directly from satellite observation data. The first of such maps contained results of observations for specific periods of time (Raschke et al., 1968). Using these data, maps of radiation regime elements averaged over several years were later constructed (Vonder Haar and Suomi, 1969; Raschke et al., 1973; Ellis and Vonder Haar, 1976).

The equation of heat balance of the earth–atmosphere system can be presented as

$$R_s = F_s + L(E - r) + B_s, \tag{3.2}$$

where R_s is the radiation balance of the earth–atmosphere system; F_s the income or outgo of heat resulting from advection in the atmosphere and hydrosphere; $L(E - r)$ the difference between heat outgo for evaporation and heat income from condensation, the latter assumed to be proportional to the rate of precipitation r; B_s the rate of heat content variation in a vertical column of the atmosphere and the upper layers of the hydrosphere (or the lithosphere).

The radiation balance is considered to be equal to $Q_s(1 - \alpha_s) - I_s$, where Q_s is the income of solar radiation at the top of the atmosphere, α_s the albedo of the earth–atmosphere system, and I_s the net longwave emission outgoing into space. The term F_s is assumed to be equal to $F_a + F_o$, where F_a is heat redistribution in the atmosphere and F_o heat redistribution in the hydrosphere.

Latitudinal means of heat balance components for the earth–atmosphere system obtained from observational data (Budyko, 1974) are listed in Table 3.3.

In this table, Q_a is the radiation absorbed by the earth–atmosphere system and I_s the longwave outgoing emission at the top of the atmosphere. The difference between these two values, R_s, is the radiation balance of the earth–atmosphere system. A total heat redistribution due to atmospheric advection equal to $F_o + L(E - r)$ is denoted by C_a. The values for

Table 3.3

Latitudinal Means of the Components of the Heat Balance of the Earth–Atmosphere System[a]

Latitude (degrees)	April–September					October–March				
	Q_a	C_a	F_o	B_s	I_s	Q_a	C_a	F_o	B_s	I_s
North										
80–90	7.8	−4.5	0	0.8	11.5	0.1	−9.1	0	−0.8	10.0
70–80	8.2	−4.4	0	0.8	11.8	0.5	−9.3	0	−0.8	10.6
60–70	11.5	−1.8	−0.4	1.2	12.5	1.8	−6.7	−1.5	−1.2	11.2
50–60	14.6	0.0	−1.3	2.8	13.1	4.0	−4.7	−0.5	−2.8	12.0
40–50	16.9	1.0	−2.0	3.9	14.0	6.5	−2.9	0.6	−3.9	12.7
30–40	19.2	1.4	−1.7	4.3	15.2	9.5	−0.9	0.7	−4.3	14.0
20–30	20.0	2.0	−0.4	2.9	15.5	13.7	0.9	0.6	−2.9	15.1
10–20	19.7	2.4	0.9	1.4	15.0	17.0	1.8	1.0	−1.4	15.6
0–10	18.4	1.4	2.2	−0.1	14.9	18.7	2.1	1.4	0.1	15.1
South										
0–10	18.0	2.3	2.2	−1.5	15.0	19.7	2.4	0.8	1.5	15.0
10–20	16.2	2.6	0.9	−2.5	15.2	20.7	3.5	−0.3	2.5	15.0
20–30	13.0	1.4	0.3	−3.4	14.7	20.6	3.5	−1.2	3.4	14.9
30–40	8.9	−0.4	−0.4	−4.2	13.9	18.9	2.0	−1.4	4.2	14.1
40–50	5.9	−1.9	−1.6	−3.5	12.9	15.9	0.0	−0.6	3.5	13.0
50–60	3.3	−3.9	−2.6	−2.5	12.3	12.8	−2.6	0.6	2.5	12.3
60–70	1.0	−9.5	0	−0.8	11.3	8.1	−4.3	0	0.8	11.6
70–80	0.2	−9.8	0	0	10.0	4.4	−6.5	0	0	10.9
80–90	0.0	−8.8	0	0	8.8	3.4	−6.9	0	0	10.3

[a] Data is for two half-years and is given in kilocalories per square centimeter month.

F_0 in Tables 3.1 and 3.3, determined by independent methods, differ but slightly. The upper left and lower right parts of Table 3.3 refer to the warm seasons, whereas the lower left and upper right parts refer to the cold seasons.

This table shows that the absorbed radiation is not the only factor determining the values of outgoing emission. In high and mid-latitudes during the cold season (and in high latitudes of the Southern Hemisphere during the whole year), the main source of heat is its transfer from lower latitudes by atmospheric circulation.

Of the two processes of heat exchange in the oceans, seasonal heat accumulation and expenditure in oceanic waters is the more important. In some latitude belts, the value of this heat balance component can reach 25–30% of the outgoing radiation. The redistribution of heat by marine currents plays a smaller role, although in particular zones its value can amount to 15–20% of the longwave radiation. It should be noted that the values of the heat balance components under consideration are related to the total areas of the latitudinal zones (in most cases, this essentially decreases their values as compared with the heat balance components related to the ocean surface areas).

As seen from Table 3.3, to determine the outgoing emission from other heat balance components one should take into account all the balance components incorporated in the table.

Table 3.4 presents latitudinal means of the radiation regime components of the earth–atmosphere system obtained by Ellis and Vonder Haar (1976) from satellite observations for two months and for the whole year.

In Fig. 3.6 one can see latitudinal means of the annual values of the earth–atmosphere radiation balance obtained from Budyko (1963) (R_s') and from the data of Ellis and Vonder Haar (1976) (R_s''), as well as similar values of outgoing longwave emission (I_s' and I_s''). As seen in this figure, calculated values of the radiation balance and outgoing radiation and those obtained directly from satellite observation data are somewhat different, because of errors in determining these values by each of these two methods. It may be assumed that the results of direct satellite observations have greater accuracy than the computed values.

However, the differences between the results of climatological calculations (Fig. 3.6, solid lines) and satellite observations (dashed lines) in many cases are so small that to speak of the advantage of one method over the other is impossible. For example, the radiation balance data obtained from observations and calculations at all latitudes except the Antarctic differ by no more than the probable error of any of the methods used for their determination. Data on outgoing emission coincide completely in high and middle latitudes of the Northern Hemisphere, where

Table 3.4

Components of the Radiation Regime of the Earth–Atmosphere System[a]

Latitude (degrees)	June			December			Year		
	α_s	I_s	R_s	α_s	I_s	R_s	α_s	I_s	R_s
North									
90–80	0.59	13.34	−0.21	—	9.56	−9.56	0.59	10.99	−6.49
80–70	0.55	13.46	0.70	—	9.76	−9.76	0.54	11.21	−5.89
70–60	0.44	14.14	0.62	0.51	10.22	−10.09	0.45	11.89	−4.53
60–50	0.40	14.57	3.29	0.51	11.12	−9.40	0.41	12.65	−2.94
50–40	0.36	15.22	4.27	0.43	12.11	−7.66	0.36	13.73	−1.31
40–30	0.29	16.40	4.90	0.36	13.60	−5.73	0.31	15.07	0.04
30–20	0.27	17.03	4.27	0.30	15.37	−3.77	0.27	16.26	1.14
20–10	0.25	16.27	4.53	0.26	16.14	−0.82	0.25	16.17	2.86
10–0	0.27	15.68	3.01	0.26	15.85	2.13	0.25	15.72	3.70
South									
0–10	0.23	17.21	0.38	0.26	15.62	4.69	0.24	16.24	3.53
10–20	0.22	17.82	−2.77	0.24	15.99	6.40	0.24	16.77	2.56
20–30	0.27	16.78	−5.38	0.25	16.46	7.04	0.25	16.52	1.38
30–40	0.34	15.11	−7.62	0.28	15.78	7.15	0.30	15.37	0.02
40–50	0.42	13.83	−9.58	0.34	14.80	6.26	0.36	14.11	−1.72
50–60	0.54	12.76	−11.25	0.41	13.90	4.86	0.43	13.01	−3.61
60–70	0.50	11.54	−11.41	0.50	13.46	2.19	0.51	11.92	−5.38
70–80	—	9.65	−9.65	0.59	13.02	0.58	0.60	10.27	−5.63
80–90	—	9.05	−9.05	0.61	12.78	0.76	0.62	9.70	−5.52

[a] Values are given in kilocalories per square centimeter month.

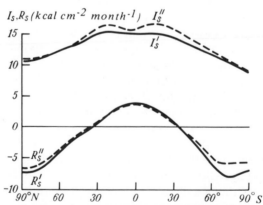

Fig. 3.6 Dependence of the earth–atmosphere system radiation balance and outgoing radiation on latitude.

more complete data from standard meteorological observations could be used in the calculation.

The differences seen for outgoing radiation in low and middle latitudes of the Southern Hemisphere seem to indicate that satellite data are more accurate in that region, where the results of standard meteorological observations are insufficient for reliable estimates. But even in this case, the discrepancy between these two kinds of data for outgoing emission is not more than 10%.

Using the available materials on the heat balance components, we can construct a diagram of the heat balance of the earth as a whole.

Considering the solar constant as 1.95 cal cm^{-2} min^{-1} and the earth's albedo close to 0.30, we find that the shortwave radiation absorbed by the earth as a planet amounts to 179 kcal cm^{-2} yr^{-1} (Q_{sa} in Fig. 3.7). Of this quantity, 118 kcal cm^{-2} yr^{-1} are absorbed at the earth's surface (Q_a) and 61 kcal cm^{-2} yr^{-1} in the atmosphere.

The radiation balance of the earth's surface is 79 kcal cm^{-2} yr^{-1} and the net longwave radiation from the earth's surface is 39 kcal cm^{-2} yr^{-1} (I). The total value of the net longwave emission from the earth, equal to the amount of absorbed shortwave radiation, is designated by I_s. The ratio of the net longwave radiation from the earth's surface to the total emission from the earth (I/I_s) is considerably less than the ratio of the amounts of absorbed shortwave radiation (Q_a/Q_{sa}). This characterizes the influence of the greenhouse effect on the radiation balance of the earth. Another characteristic of the greenhouse effect is the value of the radiation balance of the earth's surface, the radiative energy of which is spent on the evaporation of water (66 kcal cm^{-2} yr^{-1}, shown as LE) and for the sensible-heat exchange between the earth's surface and atmosphere (13 kcal cm^{-2} yr^{-1}, P).

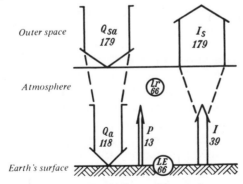

Fig. 3.7 Heat balance of the earth.

The atmosphere gains energy from three sources:

(1)   absorbed shortwave radiation, amounting to 61 kcal cm^{-2} yr^{-1};
(2)   condensation of water vapor (shown as Lr), amounting to 66 kcal cm^{-2} yr^{-1}; and
(3)   sensible-heat flux from the earth's surface, amounting to 13 kcal cm^{-2} yr^{-1}.

The sum of these three values is the output of heat as longwave radiation into space, which is the difference between the values I_s and I, i.e., 140 kcal cm^{-2} yr^{-1}.

The water balance

Data on the water balance of the earth's surface are used both in determining heat balance components and in developing the semiempirical theory of climatic change.

The water balance equation for land has the form

$$r = E + f + b, \tag{3.3}$$

where r is the precipitation, E the evaporation, f the runoff, and b the rate of change in the water content of the upper layers of the soil, all in centimeters per year. This equation can also be used in calculating the water balance of reservoirs. In this case f will be the total horizontal redistribution of water and b the variation in the level of a reservoir over the period of time under consideration. The terms in the right hand side of the equation are assumed to be positive if they characterize the loss of water. For mean annual conditions b is usually close to zero.

The water balance of the earth's surface is comprehensively studied in Drozdov et al. (1974) with accompanying detailed world maps of water balance components. In this monograph it is established that the value of precipitation for the whole earth, which is equal to that of evaporation, amounts to 113 cm yr^{-1}. This value is greater than similar values obtained earlier.

Table 3.5 gives the values of water balance components for the continents and oceans from the data of Drozdov et al. (1974). The sum of water balance components of the individual oceans is equal to the water income or outgo due to water exchange between the oceans.

Using the world maps of precipitation and evaporation we can determine the water balance components for the earth's latitudinal zones. As seen from the calculated results, in different latitudinal zones the income of water vapor into the atmosphere from evaporation might or might not exceed the loss of water vapor as precipitation. Therefore, high-pressure

Table 3.5
Water Balance of the Continents and Oceans

Continents and oceans	Precipitation (cm yr^{-1})	Evaporation (cm yr^{-1})	Runoff (cm yr^{-1})
Europe	79.0	50.7	28.3
Asia	74.0	41.6	32.4
North America	75.6	41.7	33.9
South America	160.0	91.5	68.5
Africa	74.0	58.7	15.3
Australia and Oceania	79.1	51.1	28.0
All the land	80.0	48.5	31.5
Atlantic	101.0	136.0	22.6
Pacific	146.0	151.0	8.3
Indian	132.0	142.0	8.1
Arctic	36.1	22.0	35.5
World ocean	127.0	140.0	13.0
The earth as a whole	113.0	113.0	0

zones where evaporation from the surface of the oceans is considerably greater than precipitation are one of the main sources of water vapor for the atmosphere.

Removal of this water vapor surplus occurs in the equatorial zone as well as in middle and high latitudes, where precipitation exceeds evaporation.

The difference between precipitation and evaporation in a definite region equals the difference between the income and output of water vapor in the atmosphere resulting from the horizontal motion of air. The values of this difference for many regions give an idea of the importance of water vapor transfer in the atmosphere for the precipitation regime. The dependence of the amount of precipitation on water vapor transfer is considered in investigations of atmospheric water exchange.

3.2 THE THERMAL REGIME OF THE ATMOSPHERE

The goals of the theory of climate change

The theory of climate aims at determining by the method of physical deduction the mean distribution of meteorological parameters in space and time and their variability depending on given values of external climate-forming factors. In developing a climate theory, the methods of hydromechanics and thermodynamics are used, taking into account the nature of radiative energy transfer, phase transformation of water, and other relationships of modern physical meteorology.

One of the first numerical models of climate theory was elaborated by Milankovich (1920, 1930), who calculated the distribution of latitudinal air temperature means using data on radiation income to the outer boundary of the atmosphere. He believed that the thermal regime was determined by radiative heat exchange in an individual latitudinal zone and he did not take into account the influence of the greenhouse effect and variations of albedo with latitude. He also neglected the effect of meridional heat transfer in the atmosphere and of heat exchange on the thermal regime caused by phase transformations of water and interactions between the atmosphere and ocean. Considerable errors in computing the temperature field related to the influence of these factors were compensated for by the choice of some parameters of the model. As a result, Milankovich obtained a distribution of mean latitudinal temperatures not very different from the observed ones.

Kochin (1936) was the first to apply the methods of dynamic meteorology to the development of a climate theory. He calculated the mean meridional profiles of pressure and wind speed in the lower layers of the atmosphere from a given distribution of the air temperature and pressure at the earth's surface. Afterwards, Dorodnytsin *et al.* (1939), using a similar method, constructed a model of zonal circulation for summer in the Northern Hemisphere. This kind of investigation was further developed by Blinova (1947) and others, who calculated mean temperature, pressure, and wind speed distributions in the Northern Hemisphere.

Shuleikin (1941) made a significant contribution to the investigations of climate theory. He established that ocean–atmosphere interactions are an important and often a controlling factor for climatic conditions.

Later, many new models of climate allowed the average fields of meteorological parameters to be calculated. Progress in computer engineering made it possible to develop numerical models reproducing unaveraged fields of meteorological parameters as air temperature and precipitation. As a result of the calculation of such unaveraged fields over sufficiently long periods of time, the mean distribution of meteorological elements could be computed.

Many studies in this area were conducted by Smagorinsky (1963) and his colleagues (Manabe and Bryan, 1969; Holloway and Manabe, 1971; Manabe and Wetherald, 1975; Wetherald and Manabe, 1975).

Manabe and Bryan in their paper developed a numerical model of climate theory taking into account the effect of water circulation in the oceans on climatic conditions. Holloway and Manabe (1971) were the first to construct from theory world maps of the heat and water balance components of the earth.

Various models of climate theory have been successfully developed by many other scientists (Mintz, 1965; Gates, 1975; Shvets *et al.*, 1970; Marchuk *et al.*, 1977; Meleshko *et al.*, 1979). A number of reviews of studies on climate theory have been recently published (Smagorinsky, 1974; Schneider and Dickinson, 1974; Gates, 1975, 1979).

The question of the possibility of using the theoretical models to investigate climatic change needs special study. It has been proposed that models of climate theory used for calculation of climatic change should conform to the following requirements, which are less significant for the models used in studying the present-day climate (Budyko, 1974):

(1)   The model should not include empirical data on the distribution of climate parameters, particularly those that vary considerably as the climate changes.

(2)   The model should consider realistically all the kinds of heat inflow that noticeably affect the temperature field. The model should fulfil the law of conservation of energy.

(3)   The model should take account of the principal feedbacks between different climate parameters.

The first two requirements are more or less obvious; the third will be treated more comprehensively.

The stability of the climate is provided by what can be called negative feedbacks among its parameters (SMIC, 1971). These feedbacks aid in decreasing the anomalies of meteorological elements and in changing their values toward the climatic norms. An example of a negative feedback is the dependence of longwave emission on the earth's surface temperature. With rising temperature, the longwave radiation increases, leading to an increased heat output and decreased temperature.

Of great importance for climatic changes are positive feedbacks that contribute to increasing the anomalies of meteorological elements and decreasing the stability of the climate.

The dependence between the absolute humidity of the air and its temperature, noted in Chapter 1, is one of the positive feedbacks. With rising temperature, evaporation from water or wet surfaces usually increases and this leads to a comparative stability of the relative air humidity in most climatic regions (except dry continental areas). Under these conditions the absolute air humidity increases with increasing temperature. This is corroborated by numerous empirical data.

Since net longwave radiation decreases with increasing absolute humidity, an increase in absolute humidity brings about a further increase in temperature. It is obvious that numerical models of the thermal regime used for studying climatic changes should take this feedback into account.

Another positive feedback that results from the influence of snow and ice cover on the albedo of the earth's surface is of great importance to the regularities of changes in the atmospheric thermal regime.

Voeikov (1884) was the first to investigate the climatic effect of snow cover. He established that snow cover contributes to lowering the air temperature over its surface. Later, Brooks (1950) concluded that ice cover, because of its high albedo, appreciably lowered the air temperature, with the result that, when ice formed or melted, changes in climate increased considerably.

From observational material obtained in Arctic and Antarctic conditions and from satellite data, the albedo of the earth's surface and the earth–atmosphere system in high latitudes can be estimated and then compared with that observed in regions without snow and ice cover. The available materials show that in summer the albedo of the ice surface in the central Arctic is ~ 0.70, whereas in the Antarctic it is $\sim 0.80-0.85$.

Taking into account that the average value of the albedo of the earth's surface for snow and ice free regions does not exceed 0.15, it can be concluded that snow and ice cover, other things being equal, decrease by several times the radiation absorbed at the earth's surface.

The albedo of the earth–atmosphere system is also greatly affected by snow and ice. From satellite observations (Raschke *et al.*, 1968, 1973) the albedo of the system in the central Arctic and Antarctic in summer is ~ 0.60, approximately twice the value of the planetary albedo given in these papers. It is evident that these large differences in albedo should have a pronounced effect on the atmospheric thermal regime.

If a decrease in air temperature at the earth's surface results in the formation of snow and ice cover, the absorbed radiation drops drastically, resulting in further lowering of the air temperature and enlargement of the area of snow and ice cover. The reverse result takes place when the temperature rises, if this leads to melting of the snow and ice.

Allowance for this feedback in the numerical model of the atmospheric thermal regime showed that it greatly affects the air temperature distribution near the earth's surface (Budyko, 1968a). To estimate this influence, a simple example can be given illustrating how the mean global air temperature will change if the earth's surface, with no cloud cover, is completely covered with snow and ice (Budyko, 1962b, 1971).

In this case, the earth's albedo increases considerably as compared with the existing value. This will affect the air temperature. The "effective temperature" of the earth, corresponding to its longwave emission, is proportional to $(1 - \alpha_s)^{1/4}$, where α_s is the albedo.

Therefore, the effective temperature will change as $[(1 - \alpha_s'')(1 - \alpha_s')^{-1}]^{1/4}$ when the albedo changes from α_s' to α_s''. Assuming that the actual albedo of the earth is 0.33 and that of dry snow cover 0.80, we find that for a snow-covered earth, the mean effective temperature must be reduced by ~75°.

It is possible that the reduction of the mean air temperature near the earth's surface when it is covered with ice will exceed the indicated value. First, from satellite data the albedo of the earth proved to be somewhat less than the value taken in this calculation; this would increase the value of the temperature drop due to the earth's glaciation. Secondly, at present, the mean temperature of the lower layers of the air increases essentially all over the earth's surface because of the greenhouse effect associated with the absorption of longwave emission by water vapor and CO_2 in the atmosphere. At very low temperatures this effect is not important and the formation of dense clouds that appreciably change the radiation fluxes also becomes impossible. Under these conditions, the atmosphere becomes more or less transparent to both shortwave and longwave radiation. The mean surface temperature with a transparent atmosphere is expressed approximately by the formula $[S_0(1 - \alpha_s)(4\sigma)^{-1}]^{1/4}$, where S_0 is the solar constant and σ the Stefan constant. As seen from this formula, at $\alpha_s = 0.80$ the mean temperature of the earth is 186°K or $-87°C$. Thus, if snow and ice covered the earth's surface completely, its mean temperature (presently 15°C) would be reduced by ~100°. This estimate shows what a large effect snow cover can exert on the thermal regime.

An attempt has been made (Budyko, 1961, 1962a, 1971) to assess the influence of polar sea ice on the Arctic thermal regime. With the use of data on the heat balance of the central regions of the Arctic Ocean and approximate relationships from semiempirical climate theory, it was established that polar ice lowered the mean air temperature in the central Arctic by several degrees in summer and by ~20°C in winter.

It was concluded that in the Arctic Ocean during the present epoch an ice-free regime could exist, although this regime would be unstable and could be changed by the restoration of the ice cover with a comparatively small climatic change. The same conclusion was also drawn by Rakipova (1962, 1966), Donn and Shaw (1966), and Fletcher (1966).

Thus permanent ice cover occupying even a small part of the earth's surface considerably affects the atmospheric thermal regime. This effect should therefore be taken into consideration in investigating climatic change.

Semiempirical model of the thermal regime

The difficulties involved in using models of climatic change theory in the investigation of climatic variations have not yet been completely overcome because these models usually do not fulfill the above requirements. In addition, as Smagorinsky (1974) pointed out, large expenditures of time, often exceeding the capabilities of computer centers having the fastest-acting computers, would be required to calculate the changes in climatic conditions on the basis of the models of general climate theory. In connection with this, Smagorinsky indicated the necessity of developing ways to parametrize atmospheric and oceanic circulation processes in order to estimate statistically the large-scale disturbances in this circulation by analogy with the usual methods for studying small-scale turbulent processes. This idea is used in a number of papers on climate theory that consider heat and water transfer in the atmosphere as a macroturbulent process. Taking into account the difficulties of applying more general climate theories to the study of its changes, an attempt was made in several papers to utilize for this purpose semiempirical models that could fulfil the above requirements at the expense of restricted goals for the models, extensive parametrization of large-scale atmospheric processes, and the use of empirical relationships.

The semiempirical model of the atmospheric thermal regime (Budyko, 1968a) that makes use of this direction of investigation is based on the following ideas.

In the absence of the atmosphere, the mean surface temperature is defined by the conditions of radiation equilibrium, i.e., longwave emission from the earth's surface is equal to the absorbed radiation. This may be expressed as

$$\delta \sigma T^4 = \tfrac{1}{4} S_0 (1 - \alpha_s), \tag{3.4}$$

where δ is the coefficient characterizing the difference in properties of the emitting surface and a blackbody, σ the Stephan constant, T the surface temperature, S_0 the solar constant, and α_s the mean albedo of the earth.

Assuming $\delta = 0.95$, $\sigma = 8.14 \times 10^{-11}$ cal cm^{-2} min^{-1}, and $S_0 = 1.95$ cal cm^{-2} min^{-1}, we find from (3.4) that at $\alpha_s = 0.33$, the mean temperature of the earth is 255°K or -18°C. From observational data, the mean air surface temperature is ~ 15°C. Thus, for the same value of the albedo, the atmosphere increases the mean surface air temperature by ~ 33°C as a result of the greenhouse effect, i.e., a greater transparency of the atmosphere for shortwave radiation than for longwave emission.

It should be mentioned that this estimate is relative: in the absence of the atmosphere, the planetary albedo could not be equal to the assumed value of 0.33.

At present, the mean albedo of the earth's surface is 0.14. It might be thought that before the formation of the atmosphere the earth's albedo was less than this and might be close to that of the Moon, 0.07. With the latter value of the albedo, the earth's temperature would be $\sim 3°C$.

To assess the effect of solar radiation income and albedo on the mean surface temperature under real conditions, it is necessary to know the dependence of longwave emission at the outer boundary of the atmosphere on the temperature distribution. This dependence can be established by two methods. The first is associated with the theoretical simulation of the vertical distribution of radiative fluxes, temperature, and air humidity. The second is based on empirical comparison of data on longwave radiation and the factors influencing this radiation, obtained from observations or calculations. This method was used by the author (Budyko, 1968a), who utilized the monthly means of outgoing radiation obtained by Vinnikov (1965) when preparing maps for "Atlas of the Heat Balance of the Earth" (Budyko, 1963). These materials incorporated data on outgoing radiation for 260 points evenly situated on the continents and oceans of the earth. Therefore, 3120 values were obtained.

After comparing these data with different elements of the meteorological regime, it was possible to establish that the monthly means of outgoing emission depend mainly on the air temperature near the earth's surface and on cloudiness. This dependence has the form

$$I_s = a + bT - (a_1 + b_1 T)n, \tag{3.5}$$

where I_s is the outgoing emission in kilocalories per square centimeters month, T the air temperature in degrees Celcius, n the cloudiness measured in fractions of a unit, and the coefficients are $a = 14.0$, $b = 0.14$, $a_1 = 3.0$, and $b_1 = 0.10$.

The average deviation of the values of outgoing emission obtained from monthly data by calculation using Eq. (3.5) proved to be small—less than 5%. This corroborates the conclusion that the outgoing emission is only slightly affected by other factors, vertical temperature gradient included. It should be indicated that this conclusion holds true for outgoing emission averaged for large geographical regions and over long periods of time. Physically, this conclusion means that the outgoing emission originates in the troposphere, where temperature deviations from the mean gradient are usually small compared to spatial variability or variations in the annual course.

Note that (3.5), obtained from empirical data, takes account of the previously mentioned feedback between changes in radiation, absolute humidity, and temperature.

We compare (3.5) with the results obtained in several theoretical studies, one of which, by Manabe and Wetherald (1967), is of great importance. In this paper it was established that the outgoing emission depends on temperature and on cloudiness. This dependence was presented as a graph with the lines approximately following Eq. (3.6) (with the terms and dimensions as above):

$$I_s = 14 + 0.14T - 1.6n. \tag{3.6}$$

As is seen, (3.5) and (3.6) obtained independently coincide for clear sky conditions.

Dependences similar to (3.5) and (3.6) were subsequently found from satellite observations of outgoing radiation. They are comprehensively discussed below.

To check the equations relating the outgoing emission to meteorological factors, we should use the condition of equality between the outgoing emission for the entire globe and the absorbed radiation:

$$Q_{sp}(1 - \alpha_{sp}) = I_{sp}, \tag{3.7}$$

where Q_{sp}, α_{sp}, and I_{sp} refer to the planet as a whole.

Taking into account that the value of Q_{sp} averaged over the whole earth is 20.8 kcal cm^{-2} month^{-1}, the mean air surface temperature is 15°C, and mean cloud cover is 0.50, we find that, using (3.5), the conditions of (3.7) are fulfilled at $\alpha_{sp} = 0.33$. This value of the planetary albedo differs but slightly from the estimates obtained by independent methods in a number of modern investigations, thus attesting to the validity of (3.5).

We note that in accordance with the above considerations about the effect of snow and ice cover on the radiation balance, it can be shown that (3.5) and (3.7) have at least one other solution in addition to the modern climatic regime.

Assume that at low (below freezing) temperatures near the earth's surface the entire globe is covered with snow and ice and the global albedo of the earth–atmosphere system in this case is approximately that of the central Antarctic, i.e., 0.6–0.7. From (3.5) and (3.7) it follows that the mean planetary temperature near the earth's surface varies from -47°C (at lower albedo) to -70°C (at a higher albedo). At these temperatures, complete glaciation of the earth is inevitable. This corroborates the above view that the existing external climate-forming factors can correspond not only to present-day climatic conditions but also to a climate of complete global glaciation. This question is discussed at greater length below.

In individual latitudinal zones of the earth, the thermal regime is essentially affected by horizontal heat redistribution in the atmosphere and hydrosphere. The problem of quantitative consideration of the effect of horizontal heat redistribution on the thermal regime of the atmosphere is very complex.

In studies on climate theory, it was shown that all the forms of horizontal heat transfer in the atmosphere and hydrosphere of magnitude comparable to the value of absorbed radiation should be taken into account in constructing realistic models of the thermal regime. Among these are heat transfer by regular motions and macroturbulence in the atmosphere and hydrosphere and heat redistribution by phase transformations of water. It is difficult to determine reliably the characteristics of horizontal heat redistribution in the atmosphere and hydrosphere by using the methods of dynamic meteorology. To solve this problem, we could use information on the components of the heat balance of the earth–atmosphere system.

The heat balance equation of the earth–atmosphere system has the form

$$Q_s(1 - \alpha_s) - I_s = C + B_s, \qquad (3.8)$$

where $C = F_s + L(E - r)$, i.e., the sum of heat inflows caused by horizontal motions in the atmosphere and hydrosphere.

For mean annual conditions, B_s, characterizing the accumulation or loss of heat over the time period in question, is equal to zero and the heat inflow C equals the value of the radiation balance of the earth–atmosphere system. Since the latter value can be determined either from observational data or by calculations, it is evident that the values of the horizontal heat redistribution could be found at the same time.

It might be supposed that the values of C are related in a certain way to the horizontal distribution of the mean troposheric temperature. Taking into account that air temperature deviations from the mean gradient in the troposphere are small compared to the geographical variability of temperature, we might think that the mean air temperature in the troposphere is closely allied to that at the surface of the earth. This assumption is corroborated by comparing monthly temperature means at the earth's surface and at the 500-mbar level in different geographical regions and different seasons (Kagan and Vinnikov, 1970). Thus, it may be anticipated that there is a relationship between horizontal heat transfer and temperature distribution near the earth's surface.

Since the meridional heat transfer takes place from warmer to colder regions, the value of C seems to depend on $T - T_p$, where T is the mean temperature at a given latitude and T_p the mean planetary temperature of the lower air layer. A similar assumption was made by Öpik (1953) and Sawyer (1963, 1967).

To study this relationship, Budyko (1968a) used estimates of the radiation balance of the earth–atmosphere system that were equal to values of C for mean annual conditions. The conclusion was drawn that values of R_s averaged over latitudinal zones were proportional to the differences

$T - T_p$, i.e.,

$$Q_s(1 - \alpha_s) - I_s = \beta(T - T_p), \tag{3.9}$$

where $\beta = 0.235$ kcal cm^{-2} month^{-1} deg^{-1}. This dependence simplifies the accounting of meridional heat redistribution in the model of the thermal regime.

From (3.5) and (3.9) we find that

$$T = \frac{Q_s(1 - \alpha_s) - a + a_1n + \beta T_p}{\beta + b - b_1n}. \tag{3.10}$$

With this equation we can calculate mean annual temperatures at different latitudes.

The determination of the model's parameters

The results of calculations of the atmospheric thermal regime depend strongly on the assumed values of the albedo for the earth–atmosphere system. Two values of the albedo were used in the calculation (Budyko, 1968a): 0.32 for an ice-free region and 0.62 for the Arctic polar ice region. The albedo in the polar ice zone was assumed to be constant if the area of polar ice changed. It was established that the albedo of the earth–atmosphere system is a function of the angle of incidence of solar rays: it decreases with increasing angle. But this dependence is comparatively weak when albedo values are large, i.e., for the ice-covered areas. Also, with an increase in the areas of ice-covered regions, the albedo tends to increase due to climatic changes in such regions. These factors were expected to more or less cancel each other. Therefore, we might restrict ourselves to taking into account the average albedo in a zone with ice cover, irrespective of its extent (Budyko, 1969a).

The actual values of the albedo in different latitudinal zones, obtained by Ellis and Vonder Haar (1976) from satellite observations, are shown in Fig. 3.8, where curves 1 and 2 represent albedo values for the Northern and Southern Hemispheres, respectively. The $\Delta\varphi_N$ and $\Delta\varphi_S$ latitude belts correspond to the transitional zones from ice-free to completely ice-covered regions in each of the hemispheres.

As seen in Fig. 3.8, the Antarctic ice-covered region, which far exceeds in area the Arctic ice-covered zone, has a higher value of the average albedo. This shows that in this case the influence of the altitude of the sun on the albedo is over compensated by the effects of climatic change over the ice-covered area. The use of Fig. 3.8 data on the albedo when computing modern temperature distribution presents no problems.

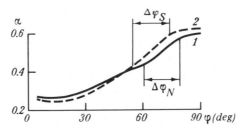

Fig. 3.8 Dependence of albedo on latitude.

The relationship between albedo values and ice position resulting from these data can be used for estimating albedo values for varying ice areas. In both hemispheres, although there is a considerable difference in polar ice area and in glaciation features, the albedo value for a zone completely covered by permanent ice is close to 0.6 (in the Southern Hemisphere the value is a little greater and in the Northern it is somewhat smaller). In transitional zones where the area is partially covered by ice, the albedo decreases as distance from the pole increases. These zones cover about 20° of latitude in each of the hemispheres.

It can be predicted that, with changes in ice area, the albedo of the earth–atmosphere system in a completely ice-covered zone would equal 0.6. The boundary of the zone is assumed to be 10° closer to the pole than the average position of the margin estimated from the area of ice.

In the transitional zones that cover 20° of latitude, the albedo can be evaluated by interpolation between the value for a completely ice-covered zone and that at the interior boundary of the transitional zone, with the assumption that the albedo of the ice-free zone does not change.

For substantiation of the albedo parametrization with variable ice area, we should consider the dependence of the latitudinal mean albedo on cloudiness. Cess (1976) presented data showing an essential increase in cloud albedo with increased latitude due to a decrease in the zenith angle of the sun. For this reason, the question arises: to what extent does an increase in albedo at high latitudes depend on the presence of ice cover and on the conditions of solar radiation reflection at the top of the clouds? In discussions of this problem, it should be kept in mind that in high latitudes, over vast ice-covered areas, there are stable regions with high pressure and low temperature and air humidity where there is usually little cloudiness. The central Antarctic at present and the regions of Quaternary glaciations in the past are examples of such climatic conditions. In this connection, a decrease in the mean albedo over a completely ice-covered zone seems to be highly improbable with the ice cover moving to lower latitudes.

It is conceivable that for a "white earth" with very low temperatures at all latitudes, the atmosphere would be practically transparent to radiation. Therefore, the earth's albedo would be close to the very high value of the albedo over clean snow cover (Budyko, 1974).

In computing the latitudinal albedo distribution, we should consider the position of the average boundary of permanent ice cover, which is dependent on thermal conditions. Since the temperature of the earth's surface differs slightly from that of the lower layer of air, it can be supposed that in the absence of the temperature seasonal march, the boundary of permanent snow and ice cover would coincide on the continents with the 0°C isotherm. This assumption may be compared with temperature estimates at the boundary of permanent snow and ice cover in different northern latitudes. The estimates were obtained from data on the height of the snow line above sea level in mountains at these latitudes, with mean values of the vertical temperature gradient taken into account. Figure 3.9 shows the results of this estimation. As is seen, this temperature actually approaches zero in the equatorial zone. In addition, in higher latitudes where an appreciable temperature variation takes place in the annual course, the snow and ice cover boundary corresponds to a negative mean annual air temperature that reaches $-11°C$ in high latitudes. Approximately the same mean annual temperature is observed at the boundary of permanent ice cover in high latitudes on the oceans.

Taking into account that continental climate conditions prevail over the ice surface, the mean annual air temperature at the ice-cover boundary in different latitudinal belts is assumed to correspond to a similar temperature at the snow boundary on the continents, as presented in Fig. 3.9.

The coefficients entering into the empirical equation (3.5) for determining longwave radiation were initially obtained from the calculation of monthly means of outgoing emission in different geographical regions. These coefficients could also be found by using satellite observations of outgoing emission (Cess, 1976). Similar computations were carried out by Vinnikov and Beeva, using more accurate data on cloud cover for different latitudinal zones. Taking their results into account, (3.5) can be represented in the form

$$I_s = 16.2 + 0.106T - 4.75n, \tag{3.11}$$

where I_s is in kilocalories per square centimeter month, T in degrees Celsius, and n the cloudiness measured in fractions of a unit.

The coefficient β in Eq. (3.9) was calculated from data on the relationship between $(T - T_p)$ (the temperature difference between individual latitudinal belts and the earth as a whole) and C in Eq. (3.8) (the value of meridional heat redistribution, equal for average annual conditions to the

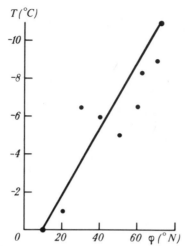

Fig. 3.9   Dependence of temperature at the snow and ice cover boundary on latitude.

radiation balance of the earth–atmosphere system as known from satellite observations). The corresponding dependence obtained from the calculated data of C by (3.11) is given in Fig. 3.10 with points depicting the values for every 10° of latitude, except for the Antarctic ice sheet, the surface of which is high above sea level. As seen from this figure, the hypothesis that values of C are proportional to those of $T - T_p$ is well corroborated by empirical data, the coefficient β being 0.232 kcal cm^{-2} yr^{-1} deg^{-1}.

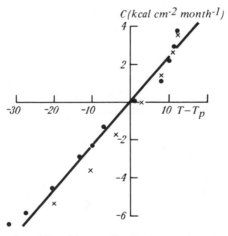

Fig. 3.10   Dependence of meridional heat redistribution on the difference between mean latitudinal and mean planetary temperatures.

This value coincides almost exactly with that obtained earlier (Budyko, 1968a).

The distribution of mean air temperatures at different latitudes can be computed by using values of the above parameters. The results of this computation are plotted in Fig. 3.11 as curve 1. Comparing them with the empirical data (curve 2) shows that the mean deviation of the measured and calculated values is 1.2°. This value is small compared with the range of variations in the mean latitudinal air temperatures, indicating a satisfactory description of real climate conditions by the model used.

Figure 3.11 does not show the values of mean air temperature at the earth's surface in central Antarctica, where this temperature is considerably lower than the results calculated using the model of the thermal regime. This difference is due to the high elevation of this region and the lack of heat transfer by oceanic currents in this zone.

The model for different seasons

To study the thermal regime of the atmosphere over different seasons, the above model should be modified, taking into account several additional factors.

A model was suggested (Budyko and Vasishcheva (1971) in which the heat-balance equation of the earth–atmosphere system is used in the form

$$Q_{sw}(1 - \alpha_{sw}) - I_{sw} = C_w + B_s, \tag{3.12}$$

$$Q_{sc}(1 - \alpha_{sc}) - I_{sc} = C_c - B_s, \tag{3.13}$$

where B_s is the gain or loss of heat due to cooling or warming of the earth–atmosphere system, determined basically by the process of cooling or

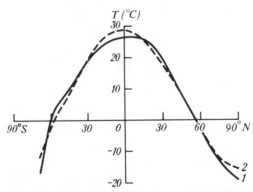

Fig. 3.11 Mean latitudinal temperature distribution.

warming of the ocean. The values referring to warm and cold seasons are indicated by subscripts w and c, respectively.

The values of I_s and C were determined by (3.5), (3.8), and (3.9), introducing an additional term characterizing a change in heat content in the earth–atmosphere system.

Equation (3.9) could be used in this case because of a correlation between the values of the meridional heat redistribution for the individual seasons, obtained from data on the heat balance of the earth–atmosphere system, and the corresponding differences in air temperature near the earth's surface. As a consequence, it was established that the dependence $C = \beta(T - T_p)$ also holds for individual seasons, the coefficient of proportionality β increasing somewhat for the cold half-year and decreasing for the warm one. This coefficient has the same value in both hemispheres in the warm period of the year. In the cold season, it is greater in the Southern Hemisphere.

The value of B_s can be found by the formula

$$B_s = s\gamma(T_{ow} - T_{oc}),\tag{3.14}$$

where T_{ow} and T_{oc} are the mean latitudinal temperatures of the surface of the ocean in the warm and cold seasons, s the ratio of the area of the oceans in a given latitudinal belt to the total area of the belt, and γ the dimensional coefficient.

For the computation of the temperature of the ocean surface, the heat-balance equation of the surface of the ocean was used in the form

$$R_{ow} = LE_w + P_w + (B_s/s),\tag{3.15}$$
$$R_{oc} = LE_c + P_c - (B_s/s) + F_o,\tag{3.16}$$

where R_o is the radiation balance of the ocean surface, LE the outgo of heat for evaporation, P the turbulent heat flux between the surface of the ocean and the atmosphere, and F_o the heat transfer by sea currents, which, as empirical data show, is mainly important in the cold season when its absolute value is comparable to that of the radiation balance.

To determine the values of LE, P, and F_o, the following approximate relationships are used:

$$LE = fT_0,\tag{3.17}$$
$$P = c(T_o - T),\tag{3.18}$$
$$F_o = \beta'(T_c - T_p),\tag{3.19}$$

where T_o is the temperature of the ocean water surface in degrees Celsius and T_c the air temperature for the cold season.

These equations are obtained by simplifying the formulas describing the heat exchange processes in the upper layers of the ocean and at its surface. Equation (3.19) was obtained by a method similar to that used to derive the equation $C = \beta(T - T_p)$.

Equation (3.17) was derived in the following way. The known equation for evaporation,

$$E = A(v)(q_0 - q) \tag{3.20}$$

(where q_0 is the specific air humidity for saturated air at the temperature of the ocean surface, q the specific air humidity, and $A(v)$ a coefficient dependent on wind speed) was somewhat simplified for approximate calculations. If the dependence of the coefficient A on wind speed v and the deviations of q/q_0 from an average value are neglected, then, using the relationship between the saturated value of humidity and the temperature, we obtain the approximate relationship (3.17).

The above formulas are simplified because in the calculation we use only values of the heat-balance components averaged over latitude and seasons and because the calculated results of the temperature distribution, as seen from data of numerical experiments, depend comparatively little on errors in determining the indicated heat-balance components.

This makes it possible to neglect the effects on evaporation of changes, for example, in relative air humidity over the oceans and in wind speed at different latitude belts, when determining the heat outgo for evaporation from the ocean surface. This simplifies the technique of calculating the air temperature and has comparatively little effect on the accuracy of the results obtained.

Numerical coefficients in these equations have the following values (for determining the heat fluxes in kilocalories per square centimeter month): $\gamma = 3.0$, $f = 0.4$, $\beta_w = 0.22$, $\beta_c = 0.27$ for the Northern Hemisphere south of the mean boundary of Arctic ice, $\beta_c = 0.40$ for the Southern Hemisphere north of the mean boundary of Antarctic ice, $\beta_c = 0.22$ for ice-covered zones, $\beta' = 0.14$ for the Northern Hemisphere, $\beta' = 0.20$ for the Southern Hemisphere, and $c = 0.84$.

From the given relationships we can derive Eqs. (3.21) and (3.22), shown on p. 87, for determining the mean latitudinal temperatures in the Northern Hemisphere.

In these and subsequent equations the subscripts to T, Q_s, n, α_s, R, and s mean that the values refer to: p, the entire planet; 1, the warm half-year in the Northern Hemisphere and the cold half-year in the Southern Hemisphere; 2, the cold half-year in the Northern Hemisphere and the warm half-year in the Southern Hemisphere. Equations (3.21) and (3.22) can be also used for calculating mean latitudinal temperatures in the Southern Hemisphere.

$$T_w = \cfrac{Q_{sw}(1-\alpha_{sw}) - a + a_1 n_w + \beta_w T_{p1} - \cfrac{\gamma s}{f+c+2\gamma}(R_{ow} - R_{oc} - \beta' T_{p2}) + \cfrac{\gamma s c}{f+c+2\gamma} - \cfrac{\gamma s(\beta'-c)(b-b_1n_w+\beta_w)}{(f+c+2\gamma)(b-b_1n_c+\beta_c)} - \cfrac{\gamma s(\beta'-c)}{f+c+2\gamma}\left[\cfrac{Q_{sw}(1-\alpha_{sw})+Q_{sc}(1-\alpha_{sc})-2a+a_1n_w+a_1n_c+\beta_wT_{p1}+\beta_cT_{p2}}{b-b_1n_c+\beta_c}\right]}{b-b_1n_w+\beta_w + \cfrac{\gamma s c}{f+c+2\gamma} - \cfrac{\gamma s(\beta'-c)(b-b_1n_w+\beta_w)}{(f+c+2\gamma)(b-b_1n_c+\beta_c)}},$$

$$(3.21)$$

$$T_c = \cfrac{Q_{sc}(1-\alpha_{sc}) - a + a_1 n_c + \beta_c T_{p2} + \cfrac{\gamma s}{f+c+2\gamma}\cdot(R_{ow} - R_{oc} + \beta' T_{p2}) + \cfrac{\gamma s(c-\beta')}{f+c+2\gamma} - \cfrac{\gamma s c(b-b_1n_c+\beta_c)}{(f+c+2\gamma)(b-b_1n_w+\beta_w)} + \cfrac{\gamma s c}{f+c+2\gamma}\left[\cfrac{Q_{sw}(1-\alpha_{sw})+Q_{sc}(1-\alpha_{sc})-2a+a_1n_w+a_1n_c+\beta_wT_{p1}+\beta_cT_{p2}}{b-b_1n_w+\beta_w}\right]}{b-b_1n_c+\beta_c + \cfrac{\gamma s(c-\beta')}{f+c+2\gamma} + \cfrac{\gamma s c(b-b_1n_c+\beta_c)}{(f+c+2\gamma)(b-b_1n_w+\beta_w)}}.$$

$$(3.22)$$

For an ice-covered zone, Equations (3.21) and (3.22) can be simplified. Taking into account that in this zone the annual variation of heat content in the earth–atmosphere system is insignificant and meridional heat transfer by currents is small or equals zero, we find that the following equations can be used:

$$T_w = \frac{Q_{sw}(1 - \alpha_{sw}) - a + a_1 n_w + \beta_w T_{p1} - lh}{b - b_1 n_w + \beta_w},$$ (3.23)

$$T_c = \frac{Q_{sc}(1 - \alpha_{sc}) - a + a_1 n_c + \beta_c T_{p2} + lh}{b - b_1 n_c + \beta_c},$$ (3.24)

where lh is the income (or outgo) of heat due to cooling (or warming) of the ice cover and freezing (or melting) of ice.

Since the value of lh is comparatively small, we restrict ourselves by considering its mean for all latitude belts, 0.8 kcal cm^{-2} month^{-1}.

To determine mean planetary temperatures, we use equations derived from (3.21) and (3.22), taking into account that for the planet as a whole the meridional heat exchange is equal to zero:

$$T_{p1} = \frac{Q_{sp1}(1 - \alpha_{sp1}) - a + a_1 n_{p1} - \dfrac{\gamma s_p}{f + c + 2\gamma}(R_{op1} - R_{op2})}{b - b_1 n_{p1} + \dfrac{\gamma s_p c}{f + c + 2\gamma} + \dfrac{\gamma s_p c(b - b_1 n_{p1})}{(f + c + 2\gamma)(b - b_1 n_{p2})}}$$
$$+ \frac{\dfrac{\gamma s_p c}{f + c + 2\gamma}\left[\dfrac{Q_{sp1}(1 - \alpha_{sp1}) + Q_{sp2}(1 - \alpha_{sp2}) - 2a + a_1 n_{p1} + a_1 n_{p2})}{b - b_1 n_{p2}}\right]}{b - b_1 n_{p1} + \dfrac{\gamma s_p c}{f + c + 2\gamma} + \dfrac{\gamma s_p c(b - b_1 n_{p1})}{(f + c + 2\gamma)(b - b_1 n_{p2})}},$$ (3.25)

$$T_{p2} = \frac{Q_{sp2}(1 - \alpha_{sp2}) - a + a_1 n_{p2} + \dfrac{\gamma s_p}{f + c + 2\gamma}(R_{op1} - R_{op2})}{b - b_1 n_{p2} + \dfrac{\gamma s_p c}{f + c + 2\gamma} + \dfrac{\gamma s_p c(b - b_1 n_{p2})}{(f + c + 2\gamma)(b - b_1 n_{p1})}}$$
$$+ \frac{\dfrac{\gamma s_p c}{f + c + 2\gamma}\left[\dfrac{Q_{sp1}(1 - \alpha_{sp1}) + Q_{sp2}(1 - \alpha_{sp2}) - 2a + a_1 n_{p1} + a_1 n_{p2}}{b - b_1 n_{p1}}\right]}{b - b_1 n_{p2} + \dfrac{\gamma s_p c}{f + c + 2\gamma} + \dfrac{\gamma s_p c(b - b_1 n_{p2})}{(f + c + 2\gamma)(b - b_1 n_{p1})}}.$$ (3.26)

By using (3.21)–(3.26) we have calculated the distribution of mean latitudinal temperatures for each season. In this calculation, it was assumed that the boundary of polar ice corresponded to the mean latitude where the temperature in the warm half-year equals $-1°C$. The albedo of the earth–atmosphere system in the polar ice zone was assumed to be 0.62 for the Northern Hemisphere and 0.72 for the Southern Hemisphere. The albedo of those latitude zones where there is no permanent snow and ice cover was found from satellite observations (Raschke *et al.*, 1968).

When determining the total radiation coming to the top of the troposphere, the radiation flux incident on an element of the surface perpendicular to a ray of the sun (meteorological solar constant) was taken as 1.92 cal cm^{-2} min^{-1}. Mean latitudinal values of the radiation balance at the surface of the ocean were taken from Budyko (1963).

The calculated results are shown in Fig. 3.12. For comparison, the figure also shows the distribution of mean latitudinal temperatures obtained from observational data. As can be seen, the discrepancy between measured and computed temperatures at different latitudes in most cases does not exceed 1 to 2°.

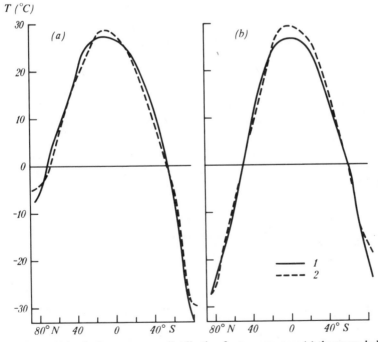

Fig. 3.12 Mean latitudinal temperature distribution for two seasons: (a) the warm half-year in the Northern Hemisphere; (b) the cold half-year in the Northern Hemisphere; curve 1, observed data; curve 2, calculated results.

3.3 THE INFLUENCE OF HEAT INCOME
ON THE THERMAL REGIME

Variations in the solar constant

Among the most general characteristics of the process of climatic change is the dependence of the mean air temperature near the earth's surface on heat inflow.

In studying this dependence, it is advisable to consider initially the dependence of the mean air temperature on solar constant variations. Chapter 4 deals with the question of variations in the astronomical solar constant. Here we mention only that, even if this value is invariable, climatic fluctuations may occur because of the variability of stratospheric transparency. This leads to variations in the so-called meteorological solar constant, i.e., the flux of solar energy coming to the upper boundary of the troposphere.

To evaluate the sensitivity of the thermal regime to variations in heat income, the parameter ΔT_1 is frequently used. It corresponds to the increase or decrease of the mean air temperature near the earth's surface produced by a 1% change in the solar constant. Table 3.6 presents values of ΔT_1 obtained by different methods.

Manabe and Wetherald (1967) were the first to evaluate realistically the sensitivity of the thermal regime to heat income variations. They took into account an important feedback among the air temperature, absolute humidity, and outgoing longwave emission, thus changing the value of ΔT_1 as compared with the previous estimates obtained without considering this feedback.

Another method of estimating ΔT_1 can be applied, based on Budyko (1968a) mentioned above, in which an empirical formula relates the value of outgoing longwave emission to the air temperature near the earth's sur-

Table 3.6
The Sensitivity of the Thermal Regime to Heat Income Variations (ΔT_1)

Parametrized models	Models of atmospheric general circulation	Materials on climate change	Satellite observations of outgoing radiation
(MW, 1967) $1.2°^a$	(WM, 1975) $1.5°$	(B, 1969a) $1.1°$	(B, 1975) $1.1–1.4°$
(B, 1968a) $1.5°$		(B, 1977b) $1.2°$	(C, 1976) $1.45°$
(B, 1979a) $1.4°$			

[a] These values are obtained in the references listed. The letters denote the authors: MW, Manabe and Wetherald; WM, Wetherald and Manabe; B, Budyko; C, Cess.

face. This equation was introduced into a numerical model of the atmospheric thermal regime that allowed us to take into account the influence on air temperature of the feedback from the polar ice position. This effect operates mainly over long-term spans; therefore, it can be neglected when computing ΔT_1 for present-day climatic changes where the duration does not exceed several decades. The ΔT_1 value obtained in Budyko (1968a) is 1.5°. When the parameters of the formula for outgoing emission were improved by using satellite data, ΔT_1 proved to be 1.4° (Budyko, 1979b).

Until recently it has not been clear to what extent we can neglect the air-temperature–cloudiness feedback when estimating the sensitivity of climate. Variations in cloudiness lead to changes in the amount of absorbed solar radiation and outgoing longwave emission. The author carried out calculations showing that this feedback is of little importance for mean global conditions (Budyko, 1971). The formula for the earth's mean temperature is used in this calculation,

$$T_\mathrm{p} = \frac{1}{b - b_1 n} [Q_\mathrm{sp}(1 - \alpha_\mathrm{sp}) - a + a_1 n], \qquad (3.27)$$

which follows from (3.5) and (3.7). This equation is solved within the limits of the actual variability of mean monthly temperatures at the earth's surface, with its accuracy depending on that of (3.5). It should be kept in mind that in accordance with the structure of (3.27), the accuracy of temperature calculations under great cloudiness (n close to 1) decreases considerably as compared to calculations for little and medium cloudiness. Although this restricts the possibilities of using (3.27), we can draw some conclusions from this formula about the effect of cloudiness on the thermal regime of the lower air layers.

For this purpose the dependence of the albedo on cloudiness should be taken into account. This dependence can be expressed as

$$\alpha_\mathrm{s} = \alpha_\mathrm{sn} n + \alpha_\mathrm{s0}(1 - n), \qquad (3.28)$$

where α_sn and α_s0 are the albedo of the earth–atmosphere system with and without cloudiness, respectively.

From (3.27) and (3.28) we have

$$T_\mathrm{p} = \frac{1}{b - b_1 n} \{Q_\mathrm{sp}[1 - \alpha_\mathrm{snp} n - \alpha_\mathrm{s0p}(1 - n)] - a + a_1 n\}. \qquad (3.29)$$

According to the technique used when constructing the maps in Budyko (1963), we assume that

$$\alpha_\mathrm{s0p} = 0.66\alpha + 0.10, \qquad (3.30)$$

where α is the earth's surface albedo. Then we find that the mean value of α_{sop} for the Northern Hemisphere is 0.20. Thus, from (3.28) we obtain that if $\alpha_{sp} = 0.33$ and $n = 0.50$, the mean value of $\alpha_{snp} = 0.46$.

Considering the values of α_{sop} and α_{snp} obtained using (3.29), we find that for the mean planetary value of Q_{sp}, the effect of cloudiness on temperature is comparatively small and seems to lie within the range of accuracy of the calculation which in this case is not high.

Schneider (1972) analyzed the effect of cloudiness on air surface temperature in more detail. He concluded that this effect depends strongly on the height of the upper boundary of the clouds. Since the relationship between the height of clouds and the air temperature has not been sufficiently studied, Schneider concluded that it is difficult to establish the sign of the effect of variations in cloudiness on air temperature.

Obviously, it is easier to solve this problem for individual regions and seasons than for mean global conditions. From the semiempirical theory of the atmospheric thermal regime, Schneider's equations, and a number of numerical models of climate theory, it follows that, for small values of solar radiation, clouds as a rule elevate the temperature near the earth's surface and, for large values of radiation, they lower this temperature.

It should be kept in mind that the cloudiness–air-temperature feedback for mean global conditions will appear with two dependences—air temperature on cloudiness, and cloudiness on air temperature.

Let us consider the second of these.

It can be seen from comparison of the distributions of cloud cover and air surface temperature that there is no simple relationship between these elements of climate. The lowest values of cloudiness are observed over the oceans and the continents, both in the region of high pressure belts covering a considerable part of the tropics and subtropics and in vast intracontinental territories in middle and high latitudes during the cold seasons. In the first case the air surface temperature is comparatively high, in the second it is very low.

Large values of cloudiness can occur both at high air temperatures (equatorial regions) and at comparatively low winter temperatures in middle latitudes in regions with marine climate.

Some dependence of the annual course of cloudiness on air temperature is evident, with large spatial averaging. Mean cloudiness for the Northern and Southern Hemispheres varies, as was shown by Berlyand and Strokina (1975), reaching a maximum during the warm season. These variations in cloudiness are, however, small and make up only several percent of its mean value.

Using the above data, it is difficult to draw a definite conclusion about the existence of a cloudiness–air-temperature dependence for mean global conditions.

Such a dependence has been recently introduced into some numerical models of climate theory, e.g., the model constructed by Paltridge (1974). Using the heat-balance equations, Paltridge obtained the dependence of the mean surface temperature on the solar constant with varying cloudiness. Since the number of exact heat-balance equations is insufficient for the determination of an additional variable—cloudiness—Paltridge added to them an equation that has no clear physical sense, which allowed the cloudiness–temperature dependence to be expressed quantitatively. Using these equations, Paltridge obtained $\Delta T_1 = 0.35°$; i.e., ΔT_1 proved to be considerably smaller than for the earth without an atmosphere.

Paltridge found that in January, when the earth is closest to the sun, the mean global surface temperature should be 2° higher than in July and the mean cloudiness should be increased by 0.05.

As seen from observed data, his conclusions were wrong. The mean air temperature near the earth's surface in January is 3° lower than in July and cloudiness in these two months is practically the same.

However, of primary importance is a check, not of these particular conclusions drawn from Paltridge's model, but of the ΔT_1 value obtained, which differs considerably from the above estimates for an atmosphere with constant cloudiness.

Weare and Snell (1974) also endeavored to take into account the air-temperature–cloudiness dependence. They calculated the thermal regime for mean global conditions, assuming that a thin cloud layer existed with properties determined by thermodynamic relationships and depending on temperature. Although the authors incorporated feedback between the thermal regime and the position of the ice and snow cover into their model, considerably increasing the climate's sensitivity to variations in the solar constant, the ΔT_1 value obtained turned out to be 0.7°; i.e., it was noticeably smaller than in the models disregarding the air-temperature–cloudiness feedback. The extremely simplified method for simulating cloud cover that was used by Weare and Snell made it difficult to estimate the accuracy of the result obtained.

This study has been further developed in papers by Temkin *et al.* (1975), Temkin and Snell (1976), and Lee and Snell (1977) that also consider the strong negative feedback between air temperature and cloudiness that appreciably decreases the sensitivity of the thermal regime to variations in heat inflow.

From our point of view, it is necessary to use empirical data (Budyko, 1975) to estimate the dependence of the mean surface temperature on radiation inflow for the real atmosphere. For this purpose we use the equation of heat balance for the terrestrial globe with solar constant S,

$$\tfrac{1}{4}S(1 - \alpha) = I_s, \tag{3.31}$$

as well as with $S + \Delta S$,

$$\tfrac{1}{4}(S + \Delta S)(1 - \alpha - \Delta\alpha) = I_s + \Delta I_s, \tag{3.32}$$

where $\Delta\alpha$ and ΔI_s are the corresponding variations in albedo and outgoing emission. From (3.31) and (3.32) we find

$$\Delta S(1 - \alpha - \Delta\alpha) - S\,\Delta\alpha = 4\,\Delta I_s. \tag{3.33}$$

We assume that within the range ΔT of air temperature change, $\Delta\alpha = A\,\Delta T$ and $\Delta I_s = B\,\Delta T$ and also that $|\Delta\alpha| \ll 1 - \alpha$. In this case,

$$\Delta T/\Delta S = (1 - \alpha)/(SA + 4B). \tag{3.34}$$

The value of B can be determined from satellite observations of the outgoing longwave emission.

Figure 3.13 shows data from Nimbus 3 (Raschke *et al.*, 1973) on outgoing emission in the Northern and Southern Hemispheres for different seasons plotted against the mean air temperature near the earth's surface for the same periods. From Fig. 3.13 it follows that the values under consideration are closely connected. This dependence is an almost linear one that justifies the above hypothesis about the dependence of outgoing emission on temperature.

From Fig. 3.13 we find that $B = 0.0024$ cal cm^{-2} min^{-1} deg^{-1}. Similar satellite observation data for 1962–1966 (Vonder Haar and Suomi, 1971) give almost the same value for B.

If the albedo of the earth–atmosphere system is 0.3 and does not depend on the air temperature ($A = 0$), then from (3.34) with B given above,

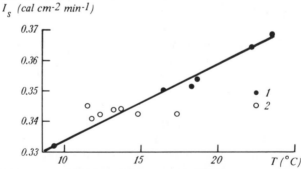

Fig. 3.13   Dependence of outgoing radiation on air temperature; 1, Northern Hemisphere; 2, Southern Hemisphere.

we find $\Delta T_1 = 1.4°$. This value agrees well with the above estimates obtained from models of the thermal regime of the atmosphere with constant cloudiness.

It is difficult to determine A from satellite observation data. Data of observations for the 1962–1966 period show that there is no correlation between the albedo and the mean air temperature for individual months in the Northern and Southern Hemispheres. This conclusion can probably be explained by insufficient accuracy of these observations.

The Nimbus 3 observation data (Raschke *et al.*, 1973) reveal some correlation between mean values of the albedo and the temperature, but the points on the diagram are too scattered to determine exactly a parameter describing this rather weak dependence.

In addition, the dependence of the albedo on the air temperature in the Northern and Southern Hemispheres as derived from satellite data may be determined largely by the annual course of radiation in high latitudes rather than by the cloudiness effect of temperature.

Since polar ice cover reflects an appreciable amount of radiation in summer, there is an increase in the albedo for the entire hemisphere during the warm season. Because of this effect, we are restricted to using satellite data for studying the dependence in question.

When evaluating the influence of temperature on the average albedo for both hemispheres, we can use the data on the annual cycle of cloudiness from Berlyand and Strokina (1975).

From (3.28) we find that the change in albedo with variations in cloud cover is

$$\Delta\alpha = (\alpha_n - \alpha_0)\,\Delta n. \tag{3.35}$$

Using this equation we can calculate the deviation of the albedo from its annual mean for different months and both hemispheres and compare them with the deviation of the average surface temperature from its annual mean for each hemisphere and for the same months. This comparison shows that the relationship between cloudiness and temperature anomalies is not close.

The value of parameter A obtained by this analysis is 0.00125 deg^{-1}; then from (3.34) we find $\Delta T_1 = 1.1°$. It should be noted that the accuracy of the determination of A by this method is not high. The conclusion that SA is considerably less than $4B$, i.e., that the cloudiness–temperature feedback does not have much effect on the dependence between the mean hemispheric air temperature and solar radiation inflow, seems to be the main result in this case.

Cess (1976) also used satellite data when studying the effects of variations in cloudiness on the dependence between air temperature and heat income. Using the outgoing emission measurements of Ellis and Vonder Haar (1976), Cess determined the empirical coefficients of (3.5) and then used this formula, together with the heat-balance equation of the earth–atmosphere system, for computing ΔT_1 values. Thus, ΔT_1 was found to be 1.45° for the Northern Hemisphere and the influence on it of the cloudiness–temperature feedback was insignificant. This feedback, as Cess showed, was also of no importance for individual latitudinal zones, except for a comparatively small region at high latitudes.

Values of ΔT_1 have recently been obtained using one of the theories of atmospheric general circulation by Wetherald and Manabe (1975). From this study, the value of ΔT_1 relating to an increase in the solar constant can be compared with the above estimates. (In this case, as opposed to that of a decrease in solar constant, the effects on ΔT_1 of the feedback between the thermal regime and the polar ice position are fairly small.) Here ΔT_1 proved to be 1.5°. Afterwards, Manabe (1976) calculated the sensitivity of the climate, taking variations in cloudiness into account in a similar model. The value of ΔT_1 proved to be slightly affected by these variations.

Similar values of ΔT_1 were obtained in a number of other studies. In particular, from data on the secular trend of the mean air temperature in the Northern Hemisphere and the direct radiation under a clear sky, ΔT_1 values were found to be 1.1–1.2° (Budyko, 1969a, 1977b). The method for determining these values is set forth in the following chapters.

We note that when these values were computed, the thermal inertia effects of the earth–atmosphere system on mean air temperature variations were not taken into account. The assumption that such effects are insignificant leads to some understating of the ΔT_1 values.

Thus, at present there are four independent approaches to estimating the sensitivity of the thermal regime to mean global conditions that give almost coinciding results, permitting the indicated estimates to be used in calculating climatic changes.

All the values of the ΔT_1 parameter included in Table 3.6 refer to cases where there is no effect from feedback between the thermal regime and the polar ice position or the feedback effect is comparatively weak.

The above data show that in this case a realistic ΔT_1 value lies within the 1.1° to 1.5° range, probably closer to 1.5°.

There are several studies in which the air-temperature–cloudiness feedback is introduced into simplified thermal models of the atmosphere, with the result that ΔT_1 decreases noticeably. As mentioned above, these models are evidently unrealistic insofar as the conclusions drawn from them are contrary to both the calculated results obtained from the most detailed models of climate theory (Manabe, 1976) and the empirical analysis of the effect of cloudiness on the thermal regime carried out using satellite observations (Budyko, 1975; Cess, 1976).

To study the regularities of climatic fluctuations in the geological past it is necessary to estimate the dependence of the mean air temperature on variations in heat income as the ice-covered area changes in accordance with thermal conditions.

As the ice-covered area grows, the albedo of the earth increases. Therefore, it is evident that $\Delta T'_1$, referring to this case, should exceed ΔT_1.

Thus, the $\Delta T'_1$ parameter derived from the semiempirical thermal model should depend on the parametrizations of latitudinal albedo distribution. Table 3.7 presents the calculated results of this dependence (Budyko, 1979b).

In Table 3.7 α_1 is the albedo over an ice-free zone, $\alpha_1 = \alpha_1(\varphi)$ denotes the variations of the albedo at various latitudes from the data of Ellis and Vonder Haar, and α_2 is the albedo over an ice-covered zone. In A and B the albedo at the ice boundary is assumed to change abruptly; in C and D it is assumed to change within a 20° wide transition zone. The ice boundary in A, B, and C corresponds to $T_0 = -11°C$; in D it varies with latitude in accordance with the relationship presented in Fig. 3.9.

Data in Table 3.7 are obtained by computing outgoing radiation using Eq. (3.11). They correspond to the case of a decreasing solar constant and refer to Northern Hemisphere conditions.

The data presented in Table 3.7 show that $\Delta T'_1$ appears to vary from 1.59 to 3.9° depending on the albedo parametrization methods used in a semiempirical model of the atmospheric thermal regime. Maximum and minimum values of this parameter apparently refer to cases of less valid assumptions of albedo values and therefore do not correspond to the conditions of a real climate. The most valid albedo parametrization is utilized

Table 3.7
The Influence of Albedo Parametrization on the Sensitivity of the
Thermal Regime to Heat Income

| | Albedo | | | | | |
	α_1	α_2	Transition zone	$T_0(\varphi)$ (°C)	$\Delta T'_1$ (°)	Δq_0 (%)
A	0.32	0.62	None	-11	3.9	2.1
B	$\alpha_1(\varphi)$	0.60	None	-11	1.90	4.3
		0.55	None	-11	1.65	5.5
C	$\alpha_1(\varphi)$	0.60	20°	-11	1.73	7.3
		0.55	20°	-11	1.59	10.3
D	$\alpha_1(\varphi)$	0.60	20°	$T_0(\varphi)$	3.3	3.2
		0.55	20°	$T_0(\varphi)$	2.0	5.0

in calculating D, in which $\Delta T_1'$ is found to be $2.0–3.3°$. Taking into account that the accuracy of the available data on the albedo in high latitudes is limited, it is unlikely that this temperature range can be narrowed on the basis of calculations from a climatic model. By using the same semiempirical thermal model of the atmosphere and taking into account the albedo dependence on the angle of incidence of solar rays, Lian and Cess (1977) have found ΔT_1 to be $1.85°$. This value seems to be somewhat underestimated because it disregards the contribution of the dependence of the albedo on climatic variations over an ice-covered region.

Cess further noted that the $\Delta T_1'$ parameter could increase considerably in the case of a decrease in heat influx caused by surface albedo variations on the continents as a result of an increase in the area of arid zones. This effect seemed to appear during the development of the Quaternary glaciations (refer to Chapter 4). The quantitative estimation of such a change in $\Delta T_1'$ presents a complicated problem because of the necessity of evaluating the state of plant cover as the climate changes. Some results of studying this effect are given below.

Because of the difficulty of determining $\Delta T_1'$ using models, the empirical evaluation of this parameter is of some interest. For this purpose data on mean air temperature variations in the Northern Hemisphere can be compared with those on CO_2 concentration fluctuations during the present time and the Neogene (Budyko, 1977b).

From this comparison it follows that $\Delta T_c'/\Delta T_c = 1.5$, where $\Delta T_c'$ is the mean air temperature variation as a result of doubling the CO_2 concentration with a corresponding change in the polar ice area and ΔT_c is the mean air temperature variation due to doubling of the CO_2 concentration with a constant polar ice area. Assuming that $\Delta T_1'/\Delta T_1 = 1.5$ also, and that $\Delta T_1 = 1.4–1.5°$, we find that $\Delta T_1' = 2.2°$. Unfortunately, the accuracy of this calculation is limited. Thus, $\Delta T_1'$ may have been $2–3°$ for the Northern Hemisphere and somewhat higher in the Southern Hemisphere.

From the available estimates of the ΔT_1 and $\Delta T_1'$ parameters, it follows that the mean air temperature at the earth's surface changes more with heat income variation than would the temperature of a blackbody in a vacuum, for which the value of a similar parameter is $\sim 0.6°C$. The studies carried out have shown that, unless this effect is considered, the physical mechanism of contemporary and past climatic changes cannot be explained.

To study climatic changes it is necessary to know, in addition to the ΔT_1 and $\Delta T_1'$ estimates, how the mean latitudinal air temperature distribution varies with warming and cooling. A diagram has been constructed (Budyko, 1974) showing the agreement between mean temperature variations at different latitudes with a 2% solar constant increase calculated by

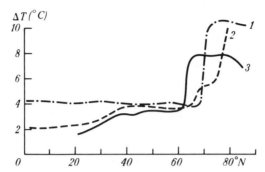

Fig. 3.14   Temperature variations at different latitudes.

a semiempirical theory of the thermal regime and by the Wetherald–Manabe model of the atmospheric general circulation. Although the agreement between results obtained by various models is very important for estimating their reliability, it is desirable to compare them with empirical data on variations in the mean latitudinal temperature distribution with climatic fluctuations.

Vinnikov and Groisman (1979) have constructed a graph (Fig. 3.14, curve 3) of the relative values of temperature variations at different latitudes, assuming $\Delta T_1 = 1.5°$. The values obtained from a semiempirical theory of the thermal regime are presented as curve 1 and values from the Wetherald–Manabe model as curve 2. As we can see, curve 3 is rather close to curves 1 and 2. Less intense global warming in high latitudes as indicated by curve 3 may be due to nonstationary heating of the earth–atmosphere system.

Calculations following different climate models and the empirical data indicate that the temperature meridional gradient increases with cooling. This conclusion, confirmed by three independent methods, is important for understanding the mechanism of the variations of atmospheric general circulation with climatic changes.

Variations in CO_2 concentration

The influence of CO_2 concentration on the atmospheric thermal regime has been studied in the past by many authors, who obtained differing results (Schneider, 1975). Results that are rather close to each other have recently been obtained using different climatic theories. Figure 3.15 shows the variations in the mean air temperature at the earth's surface at different latitudes when the CO_2 concentration is doubled. These estimates are carried out using two semiempirical models of the atmospheric

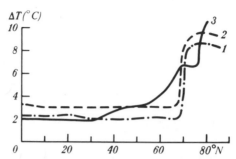

Fig. 3.15 Temperature variations at different latitudes for a CO_2 concentration twice the modern one.

thermal regime with different parametrizations of the CO_2 effect on outgoing radiation (curves 1 and 2, Budyko, 1974) and the model of atmospheric general circulation (curve 3, Manabe and Wetherald, 1975).

From the results obtained we can determine the value of ΔT_c that corresponds to an increase in the mean temperature of the lower air layers when the CO_2 concentration changes from 0.03 to 0.06%. From semiempirical models we find that $\Delta T_c = 2.5$–$3.5°$ over the Northern Hemisphere. This result can be compared with that derived from the model of atmospheric general circulation. Manabe and Wetherald (1975) have found $\Delta T_c = 2.9°$. Later they found it to be equal to $3°$ (Manabe and Wetherald, 1980).

Satisfactory agreement between these estimates and the results derived from semiempirical models does not exclude the necessity of an additional check of the ΔT_c values because of the great practical importance of the air temperature dependence on CO_2 concentration.

Although the current CO_2 concentration variations have not yet led to great climatic changes that could be easily revealed from observed meteorological data, the ΔT_c value can be estimated from data on air temperature variations over the last 100 years. Mean air temperatures in the Northern Hemisphere over the 1881–1890 and 1961–1970 periods (Budyko, 1977b) have been compared. From radiation observations it has been established that the mean atmospheric transparency during these two periods was practically the same, therefore the temperature difference in this case depended mainly on the growth of CO_2 concentration. Calculations of variations in atmospheric CO_2 concentration showed that during the second period it was 10% higher than during the first. The mean air temperature during the second period was 0.5° higher than during the first.

Taking into account that, because of the nonlinearity of the air temperature dependence on CO_2 concentration, the temperature rise amounts to

15% of ΔT_c when there is a 10% increase in CO_2 concentration, ΔT_c was found from the above data to be 3.3°.

As a result of studies of atmospheric evolution in the past, a new approach to estimating mean air temperature dependence on CO_2 concentration has recently become possible.

Using the available data on variations in atmospheric chemical composition set forth in Chapter 2, we may compare past CO_2 concentrations with the corresponding values of the mean air temperature. For this purpose, the data presented on palaeoclimatic maps compiled by Sinitsyn (1965) have been used. Since these maps cover mainly the extratropical zone in the eastern part of the Northern Hemisphere, in order to determine the temperature difference for the entire hemisphere, we have applied the method based on empirical data, showing the dependence of the relative values of mean air temperature variations on latitude.

An empirical dependence between temperature difference and CO_2 concentration is shown in Fig. 3.16. The plot is constructed from data for different periods of the Tertiary and the present time. As seen, the temperature difference corresponding to a CO_2 concentration twice the present value is 4.8°. In comparing these values, we should remember that they will differ because of temperature-field–surface-albedo feedback effects.

Since the $\Delta T_c'$ obtained from the paleoclimatic maps characterizes long-term climatic changes, it corresponds to the case when the continental ice sheets as well as the sea ice varied in accordance with thermal regime fluctuations. In addition, as the temperature increased, the albedo of the continental surface changed because of increased precipitation (see Chapters 4 and 6). As palaeogeographic data show, because of varying moisture conditions during warmer periods, dry steppe and desert areas

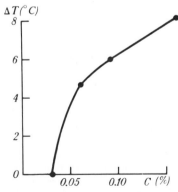

Fig. 3.16   Carbon dioxide concentration effects on air temperature during the Neogene.

decreased and forest areas increased. This led to a reduction in the continental surface albedo.

Thus $\Delta T'_c$ must exceed the ΔT_c of present-day climatic changes; then the surface albedo varies relatively little.

It should be mentioned that the difference between $\Delta T'_c$ and ΔT_c characterizing climatic changes in the past and the present, can be estimated using the following considerations.

From palaeoclimatic data it follows that during the period when the CO_2 concentration exceeded twice the present-day concentration, the area of the continental ice sheets was less than at present. This decreased the earth's albedo. In addition, during this warmer period the earth's albedo was reduced by forest expansion on the land, the albedo being less than the reflectivity of deserts and steppes.

Taking into account the results of calculating variations in the continental albedo, we can conclude that the factors indicated above had to cause an increase of $\sim 1°$ in the mean air surface temperature. If we decrease the $\Delta T'_c$ estimate given above by this value, the corresponding value of ΔT_c appears to be 3.8°.

One more method for determining the ΔT_c parameter from empirical data can be used. From the above-mentioned calculations using semiempirical climate and atmospheric general circulation models (Manabe and Wetherald, 1980), it can be concluded that the temperature rise resulting from doubling the present atmospheric CO_2 concentration is almost exactly twice the temperature elevation resulting from a 1% increase in the solar constant. Taking this into account and assuming that $\Delta T_1 = 1.45°$ obtained from satellite observations of outgoing emission (Cess, 1976), we find that $\Delta T_c = 2.9°$.

All the ΔT_c values given above are presented in Table 3.8. Because of the good agreement between all the ΔT_c values, it can be concluded that the value of ΔT_c is close to 3°.

Note that the temperature-raising effects of surface albedo variations

Table 3.8

Mean Air Temperature Variations with Doubled CO_2 Concentration (ΔT_c)

		Empirical data		
		Data on climate change		
Climate models				
Parametrized models	Models of general atmospheric circulation	Present-day	Past	Satellite observations of outgoing emission
2.5–3.5°C	3.0°C	3.3°C	3.8°C	2.9°C

due to the shrinkage of polar ice cover observed at present are taken into account when applying all the methods except the last one for determination of ΔT_c. In this connection the ΔT_c parameter for the case of constant surface albedo must be less than the value given in Table 3.8 by several tenths of a degree. Thus, the value of ΔT_c for these conditions was found to be 2.4° (Manabe and Wetherald, 1967). Augustsson and Ramanathan (1977) found that ΔT_c for different versions of the models used varied from 2.0 to 3.2°, with an average of 2.6°.

Unambiguity of climate

The question of the unambiguity of the present-day climate can be studied using the aforementioned semiempirical theory of the atmospheric thermal regime.

Lorenz (1968) investigated this question after analyzing the equations of climatic theory. He found two different solutions of these equations. The first gives one type of stable climate, by averaging the instantaneous states of meteorological element fields over long periods of time. This climate, called "transitive" by Lorenz, is the only one possible under the external conditions in question. The second solution gives several types of stable climate occurring under the same external conditions (intransitive climate).

Lorenz suggested an interesting idea that the present climate might be "almost intransitive," i.e., the climate postulated for a long-term but finite period of time could depend on initial conditions. Therefore, it could vary over different periods of time with constant external factors. The question whether the present-day climate is "almost intransitive" remains unsolved.

To clear up the question of unambiguity of climate we have used a semiempirical model of the atmospheric thermal regime and calculated the dependence of the average latitudinal boundary of polar ice cover in the Northern Hemisphere on the inflow of radiative energy at the top of the atmosphere for mean annual conditions (Budyko, 1974). This dependence is presented in Fig. 3.17, where $\Delta Q_{sp}/Q_{sp}$ denotes the relative variation in solar radiation inflow at the top of the atmosphere expressed as a percentage and φ is the average latitudinal boundary of ice cover in the Northern Hemisphere.

The relationship between φ and $\Delta Q_{sp}/Q_{sp}$ is presented as a graph that demonstrates its ambiguity and considerable variation, depending on the increase or decrease of heat inflow at the top of the atmosphere.

First we consider the case in which heat inflow increases from some initial small value. In this case, the earth is completely covered with ice ($\varphi = 0°$) that remains until the heat inflow reaches its present value (point 1) or exceeds it by 30–40% (point A). The glaciation in this case (A) is unstable and disappears with a small increase in heat income (A'). Further heat income growth maintains ice-free conditions.

When the heat inflow decreases from some initial value considerably larger than the solar constant, ice-free conditions are first observed ($\varphi = 90°$). At point E, close to the modern value of the heat income, polar glaciation begins, rapidly growing with decreasing heat income. Point 3 corresponds to the present-day climatic regime. With a decrease in heat income of 2% of its present value, the ice sheet would reach 50°N (point B). This glaciation regime is unstable and with a small reduction in heat income the earth would be completely covered with ice (B'), which would remain with further lowering of the heat income.

From the foregoing model we can also derive the dependence between φ and $\Delta Q_{sp}/Q_{sp}$ presented by the curve AB. Since this curve shows an increase in φ when $\Delta Q_{sp}/Q_{sp}$ decreases (and vice versa), we believe it may represent unstable glaciation regimes that convert into regimes of full glaciation or none with small heat income fluctuations. Therefore, point 2, denoting the present-day value of the heat income, depicts an unstable regime which could exist only temporarily.

Thus the dependence of polar glaciation on heat income has the form of a hysteresis loop with the AA' and BB' sections, designated by arrows in Fig. 3.17 representing the transition from one solution of the equations used to another. Other sections of the loop associated with stable regimes can characterize the dependence between φ and $\Delta Q_{sp}/Q_{sp}$ with both increasing and decreasing heat income.

To investigate this dependence, we can also use the model of the thermal regime for different seasons. This model gives results somewhat different quantitatively from those presented in Fig. 3.17. But qualitatively the results are similar because these qualitative regularities can be established by using a general approach, which should be considered in developing various thermal regime models.

As indicated above, with the present value of the heat income at the top of the atmosphere, a stable glaciation of the whole earth could exist, with very low temperatures at all latitudes (''white earth''). The stability of this glaciation can be attributed to the very large albedo of the surface covered with ice and snow. In this case, the climatic conditions characteristic of the Antarctic could exist over all the earth.

The conclusion of the possible existence of a stable ''white earth'' may be inferred from any realistic theory of climate. The mean temperature at the surface of the ''white earth'' can vary within a wide range depending on the albedo of the earth–atmosphere system under conditions of complete glaciation. Since this albedo can be evaluated only approximately, the available estimates of the mean temperature for the "white earth" are considerably different, but in all cases they are lower than the freezing temperature of water. This points to the great stability of the thermal regime under these conditions.

Thus the regime of complete glaciation of the earth can exist with the present value of solar radiation (point 1, Fig. 3.17). This regime is also possible if the radiation income is less than the present value (this is depicted in Fig. 3.17 by a line going along the x axis to the left of point 1) and if the radiation income is greater than the present value but less than the value that corresponds to the ice melting temperature in the warmest regions of the globe. At this temperature the ice cover over a portion of the earth's surface can melt, leading to a decreased albedo and increased absorption of radiation (A, Fig. 3.17).

We assume that between A and 3 there exist regimes of partial glaciation of the earth, which can be plotted as a line connecting these points. The form of this line can be established in the following way.

Observational data and general physical considerations show that under present-day climatic conditions, as the heat income to the earth's surface increases, the mean air surface temperature rises and polar ice recedes. As the heat income decreases, the mean air temperature drops and the ice sheets advance.

Thus the line going from A to 3 must approach the latter from the left-

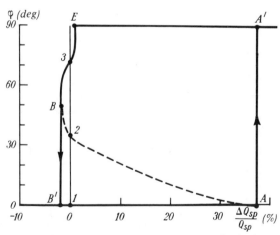

Fig. 3.17 Dependence of the mean latitude of the boundary of polar ice cover on radiation income at the top of the atmosphere.

hand side. This is possible if it cuts the y axis at least at one other point (point 2).

Therefore, we conclude that under present conditions the third climatic regime could exist as a new variant of partial glaciation of the earth with a larger area of ice coverage compared to that presently observed.

After intersecting the y axis at point 3, the line representing the regimes of partial glaciation must reach the horizontal line associated with an ice-free regime (point E). This takes place if the heat income exceeds its present value. From E the line showing further radiation increase becomes a straight horizontal line going to the right.

In addition to the dependence plotted in Fig. 3.17, the same model can be used to derive the dependence of the mean planetary air temperature on heat income, as shown in Fig. 3.18. Figures 3.17 and 3.18 use the same symbols. The dependence shown in Fig. 3.18 is similar in many respects to the corresponding dependence for the polar glaciation boundary and also has the form of a hysteresis loop.

The results shown in Figs. 3.17 and 3.18 depend essentially on the albedo parametrization of the earth–atmosphere system with and without

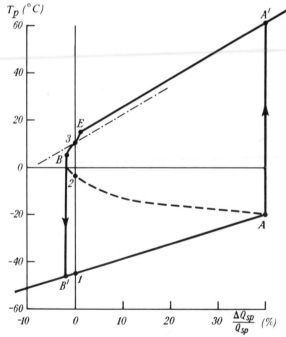

Fig. 3.18 Dependence of the mean planetary temperature on radiation income at the top of the atmosphere.

snow and ice cover. These results refer to case A in Table 3.7, from which we conclude that complete glaciation of the earth might occur with a 2% decrease in the solar constant. In other cases of the albedo parametrization presented in Table 3.7, common features of the dependences in Figs. 3.17 and 3.18 do not vary although they depend quantitatively on the albedo parametrization.

Relative estimates of a solar constant decrease (Δq_0) for various parametrizations of the latitudinal distribution of the albedo that would cause a glaciation of the earth are given in Table 3.7. These estimates lie in the 2–10% interval, whereas the most probable values (D) are 3–5%.

The problem of the earth's glaciation should be studied further to clear up incompletely solved questions.

For example, how do we explain the lack of traces of the earth's complete glaciation during Precambrian time, when the solar constant was considerably smaller than at present? Undoubtedly, at that remote epoch the atmospheric chemical composition differed greatly from the present one, significantly increasing the greenhouse effect. It is also possible that during that time, the water content of the hydrosphere was insufficient to form ice sheets over the entire planet. Although these factors explain the above contradiction, a special study of this question is needed, as discussed in Chapters 4 and 6.

A second question to be investigated concerns the possibility of the formation of ice sheets over the whole planet, given the present-day structure of the earth's surface. As palaeoclimatic data show, vast dry areas appeared on ice-free parts of the continents during the Quaternary glaciation, preventing the further advance of glaciers in several regions. We might think that in the Southern Hemisphere, which is largely covered with water, this mechanism cannot be important, and under the conditions of complete glaciation of the Southern Hemisphere, ice-free oases are unlikely to be preserved in interior regions of the Northern Hemisphere because of the drastic air temperature drop near the earth's surface. In any case, this problem should be studied in more detail.

We should emphasize that the question of the conditions that would cause the complete glaciation of the earth is not similar to that of present climate ambiguity. Thus, for example, we can prove the possibility of the existence of the "white earth," even with some uncertainties in the transitional mechanism from one type of climate to another.

The conclusion of the present-day climate ambiguity has been inferred from the thermal model describing a stationary state of the atmosphere–ocean–ice-cover system. This state can correspond to the conditions of a constant or slowly changing climate.

With comparatively drastic climatic changes, the assumption mentioned above of the stationary nature of the system may turn out to be inaccurate. Considering the great stability of the continental ice sheets in comparison with other components of this system, we can conclude that a greater number of climatic types exist under nonstationary conditions, compared with the several solutions that are associated with stationary conditions.

We assume that during the postglacial warming, large continental glaciations remained, the boundaries of which were not appropriate for the changed climatic conditions. In thise case, a new type of thermal regime appears, which can be calculated using a numerical model for a given ice-cover boundary position. The thermal regime is modified depending on the position of this boundary. As a result, an infinite number of climatic types appears, all of which agree with the given values of external climate-forming factors.

We note that present-day polar ice can be regarded as stationary since its average boundary in each hemisphere corresponds to the same values of the climatic factors that determine the sea ice position.

The stationarity in the position of the modern average boundary of polar ice in both hemispheres can be attributed to the fact that it consists mostly of marine ice of comparatively small thickness.

In Figs. 3.17 and 3.18 we can see the temperature variations which might have taken place over long periods of the earth's history. They correspond to sections of the curves going from values somewhat above point E to those above point B.

Comparison of line BE and the broken line through point 3 that represents variations in the mean planetary temperature when the polar ice exerts no effects on the thermal regime (Fig. 3.18) shows how much temperature variations due to heat inflow fluctuations are increased by polar ice effects.

Thus the present-day climate seems not to be unique for the existing climate-forming factors; i.e., it is not transitive in terms of the definition given by Lorenz. Besides the mechanism considered here, there may possibly be some other forms of climatic intransitivity, which will be established in later studies.

The results of other investigations

Following the publication of the first paper concerning a semiempirical model of the atmospheric thermal regime (Budyko, 1968a), the problem of using similar models for estimating the air temperature effects of external climate-forming factors was studied by many authors.

Sellers (1969) was the first who independently constructed a semiempirical model of the atmospheric thermal regime, based on the energy balance equation of the earth–atmosphere system with the thermal regime–polar ice feedback included. Using this model, Sellers has estimated the dependence of mean latitudinal temperatures near the earth's surface on the solar constant, the albedo of the earth–atmosphere system, and other factors influencing climatic conditions. Then Sellers used this dependence to evaluate the climatic effect of polar ice and solar radiation variations. He found that, if the albedo of the earth–atmosphere system in the central Arctic decreases from 0.62 to 0.32 when polar ice melts, the mean annual air surface temperature rises in this region by 13° and in low latitudes by 2°. He has calculated the decrease in value of the solar constant sufficient for complete glaciation of the earth to be 2% if the albedo–temperature relationship based on observational data is taken into account.

The close agreement between the results obtained by Sellers and those mentioned above (Budyko, 1970; Sellers, 1970) deserves attention, since there was some difference between these models. Sellers, in another work (1973), has developed a more detailed model of the atmospheric thermal regime for different seasons; the main results are close to those of the foregoing study.

Faegre (1972) has also constructed a semiempirical model of the atmospheric thermal regime that takes into account the thermal regime–polar ice feedback. He has also confirmed the major conclusions of the previously mentioned studies and derived from his model the conclusion that the present-day climate is ambiguous. He has found three different solutions of the equations describing the distribution of mean latitudinal temperature near the earth's surface: the first corresponds to present-day climatic conditions, the second is associated with the "white earth," and the third is intermediate between the first two.

It has been mentioned (Budyko, 1972a) that these results agree well with those inferred from our model.

Berger (1973) has shown that it also follows from Sellers' model that three climatic regimes are possible in the present epoch. The first is associated with present-day climatic conditions, the second with a colder climate (temperatures some 10–15° below the present ones), and the third with very low (below 0°C) temperatures at all latitudes ("white earth").

Schneider and Gal-Chen (1973) have studied the sensitivity of climate using semiempirical models of the thermal regime that take nonstationary processes into account. For this purpose the heat balance equation incorporates an additional term describing the thermal inertia of the system under consideration in the form

$$m \, \delta T/\delta t = \Sigma \, M_i,$$

where $\delta T/\delta t$ is the mean surface temperature variation with time, m the coefficient of the thermal inertia of the earth–atmosphere system, and ΣM_i the algebraic sum of the heat balance components of the earth–atmosphere system.

This study has confirmed the conclusion that the atmospheric thermal regime is highly sensitive to small solar constant variations and that, with unvarying solar radiation income, the thermal regime returns to its initial state, even after an appreciable decrease in the mean global temperature, to $-18°C$.

Only with a greater lowering of the mean temperature than this is the complete glaciation of the earth possible. Thus the thermal regime of the earth–atmosphere system is very stable relative to mean global temperature fluctuations that are not attributed to long-term heat income variations.

An analysis of semiempirical models of the atmospheric thermal regime that considered its nonstationary nature has been carried out by Held and Suarez (1974). They have determined the empirical parameters of the model using satellite observational data of outgoing longwave radiation and their results are similar to ours. Held and Suarez have confirmed the conclusion that the thermal regime–polar ice area feedback considerably strengthens the sensitivity of the ice cover boundary to solar constant variations and other climatic factors. They have also confirmed the existence of "critical latitudes" beyond which the polar ice expands toward the equator under unvaried external climatic conditions. The authors have considered the effects on climatic sensitivity of an additional term in the heat balance equation of the earth–atmosphere system that would characterize horizontal heat redistribution due to macroturbulent processes. This enabled them to obtain a continuous distribution of the mean latitudinal air temperature for an ice distribution concentric relative to the poles.

Su and Hsieh (1976) and Frederiksen (1976) have also studied the method for calculating the mean latitudinal air temperature distribution using this model when the albedo of the earth–atmosphere system varies jumpwise at the boundary of the ice cover. Frederiksen, using a semiempirical model of the atmospheric thermal regime, has concluded that the climate is ambiguous.

Of other studies dealing with semiempirical models of the atmospheric thermal regime, papers by North (1975a,b; North *et al.*, 1979) deserve attention. He has proposed some new methods for solving the equations of the model in question and he investigated the stability of different types of climatic conditions. North and Coakley (1978) have considered a nonstationary semiempirical model of the thermal regime.

Andronova *et al.* (1977), in solving the equations of the semiempirical theory of the thermal regime, have applied Marchuk's (1974) method of functional small disturbances, which are used for solving differential equations.

The properties of semiempirical models have been also studied by Dwyer and Peterson (1973), Gordon and Davies (1974), Hantel (1974), Chylek and Coakley (1975), and Gal-Chen and Schneider (1976).

Lindzen and Farrel (1977) believed that it could be inferred from our model that the temperature varied considerably in low latitudes during the Quaternary glaciations. This contradicts the paleoclimatic results. (They are discussed in more detail in Chapter 4.) To avoid this contradiction, Lindzen and Farrel introduced several changes into the model that produce a side effect, i.e., an increased critical value of the heat income reduction that would cause the complete glaciation of the earth. However, Lindzen and Farrel were mistaken when they believed that the conclusion drawn from our model was that the equatorial temperature varied by several degrees when the ice boundary shifted from its present position at 72°N to 60°N.

The astronomical theory of Quaternary glaciations describes a radiation decrease in high latitudes in combination with a radiation increase in low latitudes. Since Lindzen and Farrel did not consider this fact, they obtained comparatively large air temperature variations in low latitudes during the glaciations, whereas, with the correct use of our model, these changes appear to be small and similar to the paleoclimatic results.

Several papers have recently been published (Oerleman and Van der Dool, 1978; Coakley, 1979) in which the structure of the semiempirical models is not changed but larger values for the decrease in heat income that leads to complete glaciation are found as a result of a new parametrization of radiative fluxes. The second of these papers treats the problem in more detail. Coakley has established that, with the parametrization he used, global glaciation develops after an 8–9% reduction in the solar constant. A similar result has been obtained in Budyko (1979a), whose results (Table 3.7, part B) for the albedo parametrization are close to the assumptions taken by Coakley.

The above critical value of the solar constant decrease (8–9%) is probably somewhat overestimated compared with the conditions of a real climate. First, the temperature corresponding to the position of the boundary of polar ice cover increases as the latitude decreases. Secondly, when polar glaciations advance, their mean surface albedo as well as the ice–atmosphere system albedo seems to increase.

Consideration of these dependences noticeably lowers the absolute value of the decrease in the solar constant that is sufficient for the development of global glaciation. Thus, a realistic solar constant decrease that would cause complete glaciation of the earth should not exceed several percent.

A semiempirical model of the atmospheric thermal regime was developed by Gandin *et al.* (1973). They constructed a model of latitude–longitude distribution of air temperature averaged over the altitude.

By using the heat balance equation of the earth–atmosphere system, these authors have considered the horizontal heat transfer, assuming that it was determined by microturbulent exchange. The coefficient of macroturbulent heat exchange was assumed to be independent of the coordinates. The outgoing longwave radiation was estimated as a function of temperature from an empirical formula. The effect of polar ice on the atmospheric thermal regime was taken into account by use of the same method as in our model. On the basis of empirical data on the earth–atmosphere system albedo, the authors have computed the distribution of the mean annual air temperature over the Northern Hemisphere to be close to observed results.

In addition, changes in air temperature distribution and ice boundary have been studied for various values of the solar constant. It has been established that, if the solar constant increases by several tenths of a percent, polar ice melts completely, and if it decreases by the same value, complete glaciation of the earth takes place, accompanied by a drastic drop in the average global temperature ("white earth").

The authors have mentioned that the high sensitivity of the thermal regime to minor variations in the solar constant depends partly on the choice of albedo values of the earth–atmosphere system with and without ice. By decreasing the difference between these values, it is possible to obtain a critical value of the solar constant corresponding to complete glaciation of the earth that is close to the results given above. Lyapin (1977a,b) has studied this model in more detail.

Golitsyn and Mokhov have investigated the question of the sensitivity of an atmospheric thermal regime described by a semiempirical model. They used the idea of the relationship between this sensitivity and the exchange rate of entropy between the system in question and the environment (Golitsyn and Mokhov, 1978a; Mokhov and Golitsyn, 1978). This interesting idea needs to be studied further.

In their other work Golitsyn and Mokhov (1978b) showed that the sensitivity of the thermal regime varies depending on mean global cloudiness. Their estimates of the sensitivity of the thermal regime to solar constant variations make it possible to evaluate the effect of various factors on the sensitivity of the system in question. The suitability of using these estimates for the real atmosphere remains undecided because of some assumptions made in the analysis.

In spite of the approximate character of semiempirical theories of the atmospheric thermal regime, they are effective in studying the genesis of climate. In the 1970s, about 100 studies devoted to further investigation and development of these theories have been published. We believe the major merit of the semiempirical models is that, with a thorough validation of the model parameters, it is possible to fulfill the requirements, enumerated at the beginning of this chapter, to which realistic theories of climatic change must conform.

3.4   WATER EXCHANGE IN THE ATMOSPHERE

The theory of water exchange

In studies of climatic changes, along with investigations of variations in the thermal regime, it is necessary to have data on precipitation variations. To provide this, we must use the theory of water exchange in the atmosphere. This theory establishes the dependence between precipitation and the external climate-forming factors: radiation coming to the top of the atmosphere, the chemical composition of the atmosphere, and the structure of the earth's surface. Although the problem of developing a semiempirical theory of water exchange, similar to that of the atmospheric thermal regime set forth above, remains unsolved, there are several fragments of such a theory which could be used in studying climatic changes.

The water transfer equations establishing the relationship between precipitation and evaporation from the earth's surface are derived in the following way (Budyko and Drozdov, 1953). We consider the water vapor transfer in the atmosphere over a specific territory with an average linear scale L. The water vapor flux brought into this territory by air currents is wu, where w is the average water content of the atmosphere at the weather side of the border of this territory and u the mean speed of the air flux transferring water vapor over the territory. Along the direction of air flux motion, the atmospheric water content varies, depending on the difference between water expenditure as precipitation and gain from evaporation.

It is evident that the water vapor flux transferred by air currents from a given territory is equal to $wu - (r - E)L$, where r is the total precipitation, E the total evaporation for a given time period, and L the linear scale of the territory.

The total water vapor flux over the territory is composed of two fluxes: the outer (advective) flux of water vapor formed by evaporation from the areas lying outside this territory and the flux formed by local evaporation.

The first flux is wu at the weather side of the territory and is $wu - r_aL$ at the lee side of the territory (leaving the territory), where r_a is the total precipitation formed by outer (advective) water vapor. The second flux is zero at the weather side and equals $(E - r_l)L$ at the lee side (when leaving the territory), where r_l is the total precipitation formed by local water vapor.

Thus, over the territory in question, there exist average advective $wu - \frac{1}{2}r_aL$, and local, $\frac{1}{2}(E - r_l)L$, water vapor fluxes which, combined, give a total flux of $wu - \frac{1}{2}(r - E)L$. Because the molecules of local and advective vapor mix in the atmosphere due to turbulent exchange, it is evident that the ratio of total precipitation formed from local and from advective water vapor equals the ratio of the quantities of the corresponding vapor molecules in the atmosphere. Thus

$$r_a/r_l = (wu - \tfrac{1}{2}r_aL)/\tfrac{1}{2}(E - r_l)L. \qquad (3.36)$$

Therefore, with $r = r_a + r_1$,

$$r_a = r\,\frac{1}{1 + (EL/2wu)} \qquad (3.37)$$

and

$$r_1 = r\,\frac{1}{1 + (2wu/EL)}. \qquad (3.38)$$

From (3.36) we can obtain the value of the water exchange coefficient K, the ratio of the total precipitation to the advective precipitation:

$$K = r/r_a = 1 + (EL/2wu). \qquad (3.39)$$

The coefficient K shows how many times the water vapor brought into this territory condenses as precipitation before it is removed by the atmospheric circulation.

By using these formulas we can analyze the dependence of water exchange parameters on the basic factors influencing the water exchange. From (3.39) it follows that the water exchange coefficient depends on the atmospheric water vapor balance and does not depend directly on river runoff. As Eqs. (3.37)–(3.39) show, the sum of advective and local precipitation and the water exchange coefficient depend on the area of the territory in question. As it increases, the total local precipitation and the water exchange coefficient increase, and the total precipitation from advective water vapor decreases. This dependence is not linear for large areas because the average speed of water vapor transfer, u, decreases somewhat due to the effect of the curvature of the trajectory of air particle motion as the area grows.

Table 3.9 presents the calculated results of the water exchange components for the European USSR territory. Using these data we can evaluate the effect of water vapor transfer in the atmosphere on the formation of precipitation. From Table 3.9 it is seen that the precipitation formed from local water vapor makes up a small portion of the total amount. For the year as a whole and for individual months the water exchange coefficient slightly exceeds unity. This indicates that the idea of repeated interior water exchange is incorrect.

In actuality, even over the vast European USSR territory, only a small portion (about 12%) of the total precipitation consists of vapor formed by local evaporation. The major portion of the precipitation falling over a restricted land territory is formed from advective vapor.

Drozdov *et al.* (1974) has studied the atmospheric water exchange over all the continents. The results of this study are presented in Table 3.10. It can be seen that a greater portion of the precipitation is formed from advective vapor. Thus the major source of continental precipitation is moisture from the oceans.

Because of the large-scale transport of oceanic moisture in the atmosphere, the contribution of local evaporation from the surface of the continents to the total precipitation is comparatively insignificant, especially over territories less than 10^6 km^2 in area. But this does not mean that the effects of local evaporation on total precipitation are restricted by the variations in atmospheric water exchange components. In addition to the direct influence of local evaporation on water exchange there are some indirect influences due to the dependence between the amount of precipitation and the level of relative air humidity.

The factors affecting precipitation

Drozdov and Grigor'eva (1963) have derived a semiempirical formula for determining precipitation:

$$r = \alpha w \varphi(h), \tag{3.40}$$

where h is the mean relative air humidity in the 0–7 km layer, w the atmospheric water content, and α the coefficient of proportionality ($\alpha = 1$ if, at $h = 100\%$, precipitation per day is $0.1w$). For the European USSR this coefficient is close to 1.

In middle latitudes the empirical function $\varphi(h)$ has similar values for different regions; e.g., for the European USSR the $\varphi(h)$ values are presented in Table 3.11. As seen in this table, at relative air humidity equal to or less than 40%, precipitation is low and its value increases rapidly with a rise in

Table 3.9

Annual Cycle of Water Exchange Components for the European USSR

	Month													Year
	I	II	III	IV	V	VI	VII	VIII	IX	X	XI	XII	
E (mm month^{-1})	5	5	10	36	50	54	50	39	22	11	7	5	294
w (cm)	0.4	0.4	0.6	0.9	1.5	2.0	2.3	2.2	1.6	1.2	0.8	0.5	1.2
u (m sec^{-1})	7.7	7.8	7.8	7.2	6.6	6.2	5.8	6.3	6.9	7.5	7.7	7.6	7.1
K	1.07	1.08	1.09	1.24	1.22	1.19	1.17	1.12	1.08	1.05	1.05	1.06	1.12
r (cm month^{-1})	2.7	2.3	2.4	2.8	3.8	5.5	6.3	5.9	5.1	4.9	3.8	3.2	48.7
r_a (cm month^{-1})	2.5	2.1	2.2	2.3	3.1	4.6	5.4	5.3	4.7	4.7	3.6	3.0	43.5
r_l (cm month^{-1})	0.2	0.2	0.2	0.5	0.7	0.9	0.9	0.6	0.4	0.2	0.2	0.2	5.2

Table 3.10
The Atmospheric Water Exchange over the Continents

Continent	Precipitation (km³ year⁻¹)			Water exchange coefficient
	Total	Advective	Local	
Europe	7,540	5,310	2,230	1.42
Asia	33,240	18,360	14,880	1.81
Africa	21,400	15,100	6,300	1.42
North America	16,200	9,800	6,400	1.65
South America	28,400	16,900	11,500	1.68
Australia	3,470	3,040	430	1.14

air humidity up to 50–55%. With further growth in air humidity, precipitation increases more slowly. At high relative air humidity, precipitation is only slightly affected by changes in humidity.

To determine the dependence of precipitation on local evaporation we use (3.40). In this case, w and h are determined according to the effects of local evaporation on the total amount of water vapor transferred over a given territory. It can be found that the effects on precipitation of local evaporation within a definite range of air humidity can exceed by several times the contribution of local evaporation to the total amount of water vapor in the atmosphere.

Thus the theory of atmospheric water exchange enables us to consider only the direct effects of local evaporation on the amount of precipitation. As Drozdov showed, in many cases the indirect effects of local evaporation on precipitation are of great importance; without them, interior regions away from the oceans could turn into deserts.

By using (3.40) we can calculate how precipitation depends on the type of vegetation cover, which determines the quantity of local evaporation. But the dependence between total evaporation and the kind of vegetation has as yet been insufficiently studied. For instance, the effect of forests on evaporation has been discussed for many years but this question still remains unsolved. Forests seem to affect the total evaporation only slightly.

Draining and afforesting marshes overgrown with moss can increase total evaporation. The reverse effect may take place on draining low-lying grass marshes.

Table 3.11
The $\varphi(h)$ Function

h (%)	40	45	50	55	60	65	70	75	80	85	90	95	100
$\varphi(h) \cdot 10^{-2}$	0.7	1.4	3.7	5.7	6.8	7.5	8.2	8.6	9.1	9.5	9.7	9.9	10.0

Let us assume that with the changing of various types of vegetation cover the total evaporation varies by an amount of the order of 10%. Using the water exchange equation, we find that the effect of variations in evaporation on precipitation depends on the area of the region over which evaporation varies. If the change occurs within a geographic zone with a linear scale of the order of 10^3 km, precipitation is slightly affected by variations in evaporation. However, the effect may become significant if evaporation varies over all or the greater part of a continent.

The aforementioned considerations can be amplified by analyzing the physical mechanism of changes in precipitation as the air temperature field varies (Budyko and Drozdov, 1976).

We assume that evaporation from the earth's surface is

$$E = A(q_s - q),\qquad(3.41)$$

where A is the coefficient of proportionality; q_s the humidity of the air, saturated by water vapor at the temperature θ_w of the evaporating surface; and q the specific air humidity at the level of meteorological observations.

This formula is accurate for evaporation from the water surface, which makes up more than 85% of the total evaporation from the entire earth's surface, and it is approximate for evaporation from humid land surface.

Taking into account that the relative air humidity varies comparatively little over the oceans, let the value of $1 - (q/q_s) = \beta$ be assumed constant. In this case, we find the ratio of $\partial E/\partial\theta$ (the derivative of evaporation with respect to air temperature) and E (evaporation):

$$\frac{1}{E}\frac{\partial E}{\partial\theta} = \frac{1}{q_s}\frac{\partial q_s}{\partial\theta_w}\frac{\partial\theta_w}{\partial\theta}.\qquad(3.42)$$

In order to determine the value of $\partial\theta_w/\partial\theta$ we use the heat balance equation of the earth's surface for the whole earth,

$$R = LE + P,\qquad(3.43)$$

where R is the radiative balance, L the latent evaporation heat, and P the turbulent heat flux between the earth's surface and the atmosphere. This equation can be rewritten as

$$R = LA(q_s - q) + c_pA(\theta_w - \theta),\qquad(3.44)$$

where c_p is the specific heat of the atmosphere at constant pressure. Differentiating (3.44) with respect to θ, we find that

$$\frac{\partial\theta_w}{\partial\theta} = \frac{1}{1 + (L/c_p)\beta(\partial q_s/\partial\theta_w)}.\qquad(3.45)$$

In this case, it is assumed that while temperature varies, the radiative balance of the earth's surface remains constant; actually, it varies slightly.

From (3.42) and (3.45) we have

$$\frac{1}{E}\frac{\partial E}{\partial \theta} = \left(\frac{1}{q_s}\frac{\partial q_s}{\partial \theta_w}\right)\bigg/\left(1 + \frac{L}{c_p}\beta\frac{\partial q_s}{\partial \theta_w}\right). \tag{3.46}$$

Using (3.46) we can calculate the dependence of evaporation on mean global air temperature. From observational data we assume $\beta = 0.2$, and, using (3.46), find that a 1° increase in mean air temperature causes a 4% increase in evaporation. Because total global precipitation is equal to the total evaporation, this dependence also characterizes the relationship between the total precipitation and air temperature.

The dependence derived here can be compared with the results obtained by Wetherald and Manabe (1975). They used a numerical climatic model that considered the atmospheric general circulation. This approach enabled them to avoid using some simplifications in the above computation. In particular, Wetherald and Manabe did not consider the relative air humidity to be constant when temperature varied. They found that global evaporation (and precipitation) changes by 3% when temperature changes by 1°. This result does not differ greatly from that obtained using a simplified calculation.

It should be noted that the assumption that β is independent of temperature over the continents is not accurate, becase the potential evapotranspiration from land increases with a rise in temperature. In this case, if the inflow of moisture from the oceans remains constant, the mean moisture content of the upper soil layer decreases with the increase of evaporation. This results in a reduction of relative air humidity. Therefore, warming would lead to reduced precipitation over a considerable part of the continents. Brückner (1890) was the first to detect this effect. Further studies have found a negative correlation between temperature and precipitation in the warm half-year for the Eurasian continent (Drozdov and Grigor-'eva, 1963) and in various seasons over the greater part of the US territory (CIAP, 1975).

The physical mechanism of a decrease in precipitation due to an increase in the mean air temperature for the Eurasian continent can be explained as follows.

The flux of water vapor transported from the Atlantic ocean and adjoining seas to the Eurasian continent is proportional to vq_0, where v is the rate of water vapor transfer as a function of meridional temperature gradient and q_0 the air humidity at the boundary of the ocean. The equation of water vapor balance in the atmosphere over the continent at a distance x from its western coast has the form

$$avq_0 - avq = \int_0^x (r - E)\, dx, \tag{3.47}$$

where a is the coefficient of proportionality.

The values of r and E vary along the x coordinate, decreasing as the distance from the ocean increases. To determine these two values, two additional relationships can be used. One of these is an empirical dependence of precipitation on air humidity from (3.40),

$$r = r(q), \tag{3.48}$$

and the other expresses a relationship between the heat and water balances for land (Budyko, 1956)

$$\frac{E}{r} = \Phi\left(\frac{R}{Lr}\right). \tag{3.49}$$

From these equations we can find the value of r for an interior region located at a distance x from the shore.

Considering that $w = bq$ (where b is a proportionality factor), we derive from (3.40) and (3.47)–(3.49) that

$$\frac{av}{\alpha b}\left[\frac{r_0}{\varphi(h_0)} - \frac{r}{\varphi(h)}\right] = \int_0^x r\left[1 - \Phi\left(\frac{R}{Lr}\right)\right] dx, \tag{3.50}$$

where r_0 and h_0 are the precipitation and relative humidity at the boundary of the ocean. Thus precipitation over the interior regions of a continent proved to be dependent on horizontal fluxes of moisture, which were first detected empirically (Burtsev, 1955).

In order to employ (3.50), it is necessary to take into account the dependence of precipitation r on atmospheric heat transfer over a continent. In this connection, the role of local evaporation in the formation of precipitation differs depending on whether air masses from the ocean are heated or cooled over a continent. Therefore, the schemes of water exchange are different for different seasons.

Furthermore, it should be said here that the foregoing mechanisms for the formation of precipitation at middle latitudes cannot explain variations in precipitation at lower latitudes, as well as a number of distinguishing features characteristic of moistening conditions in the geological past.

Some conclusions concerning the dependence of precipitation on the main precipitation forming factors can be derived from the analysis of (3.40).

Previously we considered the implications of variations in the parameters representing the atmospheric circulation α and water content w related to temperature and moisture fluxes from the ocean at a constant level of condensation determined by relative humidity h. However, the factors in (3.40) can be assumed to be independent only conventionally.

Several types of water exchange can be described, considering the effect of air-mass heat transformation on precipitation and the relations between the factors in (3.40).

The first is water exchange when heat transformation has no effect on precipitation.

The second is observed in summer at middle latitudes when, in air fluxes directed from the ocean to the continent, the relative humidity h drops due to heating of the continent as the distance from the ocean increases, but at the same time, the index α increases. In consequence, for instance, when maritime air masses from middle latitudes or Arctic air masses intrude into the continent, heavy precipitation occurs along the coast with a decreasing amount over interior regions. This decrease, however, is compensated for by an increase in relative humidity h due to local evaporation of moisture accumulated during the cold season. Under these conditions, local evaporation is of great importance for moistening the interior regions of the continent.

The third type of water exchange takes place in winter when, because of chilling, the relative humidity in air fluxes inside the continent remains constantly high (in spite of the fall of precipitation) and the role of local evaporation in the formation of precipitation becomes secondary. This type of water exchange is typical of polar latitudes and the interior regions of large continents in middle latitudes. In central Antarctica a greater portion of the precipitation can be attributed to the freezing out of moisture, not to frontal processes (Rusin, 1961). Similar conditions are possible in winter in the region lying behind Lake Baikal (the so called Zabaikal'e) and in the Yakutsk region of Siberia.

Let us consider a particular case of using (3.50), when the air flux is from the ocean to the land and the relative humidity h and the ratio E/r do not change, i.e., $E = cr$, $\varphi(h) = $ constant. These conditions might take place as air masses over the continent cool down, i.e., when a 10° temperature decrease reduces precipitation by nearly one half.

In this case, assuming v to be constant and differentiating both parts of (3.50) with respect to x, we obtain

$$-\frac{dr}{dx} = \frac{\alpha b(1 - c)\varphi(h)}{av} r, \qquad (3.51)$$

whence, assuming that for $x = 0$, $r = r_0$, we find

$$r = r_0 \exp - \frac{\alpha b(1 - c)\varphi(h)x}{av}. \tag{3.52}$$

From (3.52) it is seen that precipitation decreases toward the interior of the continent, with the rate dependent on v. Thus variations in v induce inland variations in r, similar in sign. More complicated but similar relationships are obtained with other types of heat transformation.

A distinctive type of water exchange appears in the intertropical zone in the absence of intensive descending motions. Favorable conditions for convective precipitation are created there at high relative humidity h and water content w, since in this case the moist-adiabatic gradient decreases and in the upper layers of the troposphere radiation increases, causing a rise in temperature lapse rate. As a result, although air-mass precipitation is usually less important at middle latitudes (except in summer in some monsoon regions), it becomes most significant in the intertropical zone.

The theory of water exchange in the atmosphere developed here might be used for studying precipitation variations due to natural and man-made climatic change.

Chapter

4

Natural Climatic Changes

4.1 CLIMATES IN THE GEOLOGICAL PAST

Methods for studying past climates

Information on the climates of remote epochs has been obtained by the examination of data concerning natural conditions in the past. Since the processes of sediment formation, rock weathering, and formation of water reservoirs, and the existence of living organisms were dependent on atmospheric factors, data relating to these processes allow us to evaluate the climatic conditions of the relevant periods of time.

The interpretation of data on the natural conditions in the geological past for the estimation of a climate regime presents great difficulties, some of them crucial. Among these difficulties is, in particular, the necessity of utilizing the principle of uniformitarianism. In this case it corresponds to the assumption that, in the past, relationships between the climate and other natural phenomena were the same as at present. Although this approach is open to argument, the variety of natural processes that depend

123

on climate enables the results of past climate reconstruction to be verified independently with the help of various paleogeographic indicators. The most general regularities of climatic conditions in the geological past, as established in paleogeographic investigations, are undoubtedly reliable, whereas some individual results are open to question and require further examination.

In addition to paleogeographic data, information on paleotemperatures obtained by isotopic analyses of organic remains appears to be very helpful in the study of ancient climates. The evaluation of the accuracy and the proper interpretation of these data is a matter of some difficulty, which is gradually being overcome with improvement in the methods for paleotemperature studies.

In studying climates of the past, information on sedimentary phenomena, geomorphological evidence, and data on fossil flora and fauna are used.

The first approach is based on the well-known relations between lithogenesis and climatic factors. For example, the chemical weathering of rocks by the decomposition of unstable minerals is especially promoted under the conditions of a hot, humid climate. When the climate is hot and dry, chemical weathering is less intensive and rocks are weathered largely by wind and temperature fluctuations. Chemical weathering is even more retarded in a cold climate where physical weathering, which does not affect chemically unstable minerals, dominates.

The extent and structure of deposits are strongly dependent on moisture conditions. The volume of sedimentary deposits is usually small in dry areas and increases in humid regions, where alluvial deposits prevail.

Since coal formations are closely related to climatic conditions, data on fossil coal can be used for reconstructing past climates. However, the interpretation of these data is fraught with certain difficulties because, at different periods, the relationship between coal accumulation and climate varied to a great extent, depending on the nature of the vegetation that formed the coal. For example, many coal beds of the Devonian were formed in a dry climate, while the coal formations of the Carboniferous period were usually connected with more humid climatic conditions.

Data on deposits of limestone and dolomites, as well as on the salt deposits that are so essential for reconstructing the climatic conditions of ancient water reservoirs, are also used in studying climates of the past.

In some cases, data on the sediment structure are used for estimating seasonal variations in the climate. For example, varved clays around the margins of continental glaciations are examined for this purpose. During the summer melting of glaciers, water streams carried away a great deal of rough debris, while in cold seasons, far less fine-grained argillaceous substance was deposited in the same area. The stratified structure of varved clays allows the length of their formation period to be evaluated.

Geomorphological evidence is widely used for studying the atmospheric precipitation regime and ice cover formation. For instance, data on the position of the ocean coastline enable us to estimate the water lost or gained by the ocean in the formation or melting of continental glaciations.

Surface relief variations due to glaciations are a major indicator of glacier development. In addition, an important evidence of past climates seems to be data on snow levels in the mountains, which depend on the temperature and precipitation regimes.

Data on ancient lakes and river valleys may give certain information concerning moisture conditions. For instance, impressions of numerous lakes and rivers in present deserts are indicative of great changes in the moisture conditions in these regions.

Information on the sea-level fluctuations of such land-locked reservoirs as the Caspian Sea allows us to evaluate the water inflow and, hence the precipitation over the basins of the feeding rivers at different periods of time.

Data on fossil soils are utilized for studying moisture and thermal conditions in past times. Traces of permafrost are valuable in reconstructing zones of cold climate.

Information on the nature of erosion processes strongly dependent on climatic conditions, in particular, on moisture conditions, is also involved in paleoclimatic studies.

The geographic distribution of living organisms, particularly of vegetation vitally dependent on climatic conditions, can be of great importance in the study of past climates. This method provides quite reliable results for the not-so-distant past when plants were similar to modern species and their distribution was evidently affected by climatic conditions in much the same way as at present. Extrapolation of the relationships between modern plants and meteorological factors to more ancient times presents certain difficulties, which restrict the use of data on plant distribution in paleoclimatic studies.

Great advances have been achieved in recent decades following the development of fossil pollen and spore analysis, which allows the composition of a regional vegetation cover to be defined. This method has proved efficient in the reconstruction of the epochs when plants were similar to modern species.

We should mention here data on annual tree rings, which are also used in the study of climate change. Variations in the annual growth layers of a tree reveal short-term climatic fluctuations, whereas their structure is indicative of general climatic conditions (for example, trees growing in carbonaceous marshes exhibit poor formation of annual rings, reflecting no seasonal climate variations).

Information on fossil fauna is more difficult to regard as diagnostic of past climates, since the geographic distribution of animals is generally not as strongly dependent on climate as that of vegetation. However, these data are a valuable addition to other methods used in the study of climate change. Such information is mostly data on poikilotherms, i.e., animals without heat regulation. As was the case with plants, the data on the distribution of animals are interpreted with the assumption that climatic effects on the vital activities of the relevant organisms were similar to those existing at present.

A paleotemperature method based on the ^{18}O isotope content of the fossil remains of aquatic animals is of particular importance in paleoclimatology. It has been established that the $^{18}O/^{16}O$ isotope ratio in mollusc shells and other remnants of marine organisms depends on the temperature at which the animals lived. In order to determine paleotemperatures by this method, it was necessary to develop highly sensitive mass spectrometers and solve other technological problems. Since the middle of this century the method of direct measurement of paleotemperatures was widely applied in the studies of climatic conditions of the recent past and of hundreds of millions of years ago.

In evaluating the results of the above methods for studying climatic variations, it should be recognized that, except for a relatively short contemporary period, almost all the available information on past climates concerns the air temperature regime near the earth's surface, the surface temperature of the land, the temperature of water reservoirs, and, to a lesser degree, the moisture conditions on the continents.

Paleogeographic data can also supply information on some other climatic elements. For example, in some cases the shape of fossil dunes and barkhans can be indicative of the direction of the prevailing winds. But these data are not so ample as the above-mentioned.

Pre-Quaternary climates

The information concerning climates in pre-Quaternary time that is set forth below is based generally on Sinitsyn's investigations (1965, 1967).

Paleozoic climatic conditions (570–230 million years ago) are only vaguely known. Apparently, during much of the Paleozoic, the climate was very warm over the entire globe and moisture conditions on the continents varied over a wide range. In the late Paleozoic, i.e., at the threshold of the Carboniferous and the Permian, a glaciation developed that expanded over vast areas of land, mainly in the present tropic latitudes. It is rather difficult to estimate the geographic position of this glaciation at the time of its development, because the earth's poles and continents have changed their position greatly since then. In addition, climatic conditions in other regions of the globe were rather warm during the epoch of Permo-Carboniferous glaciation.

In the Permian the thermal zonality became more pronounced and an arid climate expanded widely over the continents.

The Mesozoic climate (230–65 million years ago) was rather equable. Throughout most of the globe climatic conditions were similar to the contemporary tropics, although in high latitudes the climate was cooler but still quite warm, with insignificant seasonal temperature variations. In the Mesozoic, moisture conditions were apparently more uniform as compared to the contemporary epoch. At the same time there were also zones of insufficient and excessive humidity.

At the end of the Cretaceous the hot climate zone decreased in size and arid climatic conditions expanded. Climatic change did not mark the transition to the Cenozoic. During the Tertiary the process of cooling, which was more pronounced in middle and especially in high latitudes, continued. In the mid-Tertiary a new climatic zone appeared and expanded gradually. Its meteorological regime was similar to modern climatic conditions of the middle latitudes. The air temperature there dropped below zero in winter, producing conditions favorable for a seasonal snow cover. At the same time, the continental climate became more extreme in the areas far from the ocean.

The process of cooling was not constant; during some periods it gave way to warming. But the general tendency toward thermal zonality, which was determined by a temperature drop in the high latitudes, remained invariable. It became more pronounced in the Neogene with the advancement of the Antarctic continental glaciation.

At the end of the Pliocene the climate was warmer than at present, yet it resembled present climatic conditions more than the climates of the Mesozoic and early Tertiary.

In examining the sequence of climatic variations throughout the period covered by more or less reliable data we should recognize that climatic regimes with strongly pronounced thermal zonality were unusual for our planet.

A great temperature difference between the equator and the poles existed since the end of the Tertiary, the gap increasing in the ice ages. But that difference is characteristic of only a short span of the time that elapsed since the beginning of the Paleozoic. Except for the Neogene glaciations, there was only one major ice advancement over the last 600 million years. It took place in the Permo-Carboniferous and also lasted a fairly short period of time compared to that which passed since the onset of the Paleozoic.

Climate of the Quaternary age

The climatic conditions of Quaternary time have been studied more thoroughly than any earlier climates. Among the numerous works devoted to natural conditions and climates of the Quaternary we mention here studies by Gerasimov and Markov (1939), Schwarzbach (1950, 1968), Saks (1953), Markov (1955, 1960), Emiliani (1955), Ewing and Donn (1956, 1958), Flint (1957), Zeuner (1959), Rukhin (1962), Flohn (1963, 1964), Butzer (1964), Fairbridge (1967), and Velichko (1973).

The entire Quaternary period, except for the last, relatively short interval called the Holocene, corresponds to the Pleistocene. During the Pleistocene climatic conditions differed greatly from the preceding Mesozoic and Tertiary when thermal zones were relatively indistinct. A cooling trend became more pronounced in middle and high latitudes and great continental glaciations developed. They advanced repeatedly, reaching the middle latitudes, and retreated again to the high latitudinal belt. During the advancement of continental glaciations, sea ice expanded over vast areas. We do not know the exact number or dates of the Quaternary glaciations.

Alpine investigations carried out in the last century gave evidence of four major European glaciations called Günz, Mindel, Riss, and Würm. Later it was discovered that each of these glaciations proceeded in several stages, with intervals marked by recession of the glacier. The Alpine glaciations were accompanied by the development of ice cover in the lowlands of Eurasia and North America, stretching over vast territories in high and middle latitudes.

The expansion and contraction of the glaciers seem to have occupied only short periods of time compared to the entire length of the Pleistocene. The interglacial epochs, which lasted longer, were relatively warm. During that time ice cover on the continents disappeared, persisting only in mountainous regions and high latitudes.

The advancement and retreat of glaciers took place more or less simultaneously in Europe, Asia, and North America. This also applies to the ice ages in the Northern and Southern Hemispheres.

During the ice ages the continental ice cover spread farther in the zones of humid maritime climate. Under the relatively dry climatic conditions of

Northern Asia it covered comparatively small areas. During the greatest glaciations the continental ice cover in the Northern Hemisphere reached 57°N on the average and even 40°N in some places. Its thickness in most areas was hundreds of meters and sometimes more than a kilometer.

Undoubtedly the progress of continental glaciation was accompanied by a shift of the polar sea-ice boundary to lower latitudes. This considerably increased the total extent of permanent ice cover on the planet.

At each glacier advancement the snow line in mountain regions outside the glaciation area dropped by hundreds of meters, sometimes by a kilometer or more. At the same time the permafrost zone increased considerably during the ice ages. With the development of glaciations the permafrost boundary moved a few thousand kilometers into lower latitudes.

The ice ages were marked by a 100–150 m drop in the level of the World Ocean, as compared to its modern level, as a result of the development of the thick glaciations on the continents. During the warm interglacial epochs the sea level was several dozen meters higher than that at present. The extent of the ocean surface varied by a few percent according to fluctuations in its level.

A noticeable drop in the air temperature occurred over the entire globe during the ice ages. In the warm interglacial epochs the air temperature was higher than at present. The problem of variations in the precipitation regime associated with glaciations is less clear. According to the available data, moisture conditions in various parts of the earth varied in different ways under glaciation conditions. This means that the atmospheric circulation changed as a result of variation in the temperature differences between the equator and the poles.

At the same time precipitation decreased in many continental regions during the epochs of the greatest glacier advances. Evidently this is accounted for by a reduction in evaporation as a result of the temperature drop and the fact that the ocean surface was partially covered with ice.

Reasons for climatic change

To understand the physical mechanism of the present changes in climate it is essential to know the causes of climatic change in the geological past.

A gradual increase in the luminosity of the sun (the solar constant) was one of the factors determining long-term climatic change. This conclusion made by Schwarzschield (1958) was based on the analysis of general regularities in the evolution of stars. It was confirmed by the results of subsequent studies but the rate of the increase in solar luminosity is not definitely known. Different authors suggest that during the earth's history (about 4.5 billion years) the luminosity of the sun increased by 25 to 60%.

Among other factors that determined climate in the past we should mention variations in the chemical composition of the atmosphere and, in particular, the atmospheric carbon dioxide fluctuations that are treated in Chapter 2.

If we took into consideration only the first of the above factors in climatic change, we should come to the conclusion that the air temperature near the earth's surface was increasing during the entire history of the earth. As seen in the above survey, this contradicts the available paleoclimatic data.

Interesting discussions of the problem are found in Sagan and Mullen (1972) and Sagan (1977). The first paper presents calculations of variations in the mean temperature of the earth's surface due to the increasing luminosity of the sun, based on the formula of the radiation equilibrium ignoring the dependence of longwave radiation on the properties of the atmosphere. It was also assumed that the earth's albedo was constant through time.

According to these calculations, the mean temperature should have been below the freezing point of water during most of Precambrian time, presumably until 1.4–2.3 billion years ago (the dating depends on the assumed rate of the solar luminosity increase). However, according to the available data, there was liquid water on the earth's surface throughout the first half of the earth's history, contradicting this conclusion. Therefore, Sagan and Mullen have assumed that the ancient atmosphere contained some ammonia (NH_3) that intensified the greenhouse effect and resulted in appreciable increase in the air temperature.

Subsequently, considering this supposition, Sagan sets forth the findings on paleotemperatures of the Precambrian period. According to these data, some 1–3 billion years ago the temperature at the earth's surface was 310–350 K, i.e., higher than at present.

The semiempirical theory of the thermal regime of the atmosphere presented in Chapter 3 adds significantly to Sagan and Mullen's arguments supporting the idea that the chemical composition of the atmosphere was quite different in the past. Their date for the appearance of liquid water on the earth's surface due to the solar constant increase should be altered. The solar constant need be only 4% less than its present value for the "white earth" to arise. This was established in the above calculations based on the semiempirical theory of the thermal regime of the atmosphere. In the early or middle Phanerozoic, i.e., 600–300 million years ago, the solar constant corresponded to just this value.

We should like to emphasize here that according to the semiempirical theory, the state of the "white earth" holds for the present-day value of the solar constant. In this connection, the earth's surface, which was frozen in the past, would be covered by ice in our time if the chemical composition of the atmosphere had remained constant.

The warm climatic conditions at all latitudes for the greater part of the earth's history are probably explained by a much higher CO_2 content in the atmosphere through the entire Phanerozoic (see Chapter 2) than today. As mentioned in Chapter 3, a doubling of CO_2 content results in an $\sim 2.5°$ increase in the mean air temperature near the earth's surface at a constant earth albedo. With a further increase of the CO_2 concentration, the air temperature will rise by the same amount at each successive doubling of the concentration (Augustsson and Ramanathan, 1977). Therefore, a fourfold increase in the CO_2 concentration adds 5° to the air temperature, and if the concentration is eight times higher, the temperature will rise by 7.5°.

In the early and mid-Phanerozoic the CO_2 concentration was 6 to 10 times higher than at present. This ensured an $\sim 6.5-8°$ increase in the mean air temperature as compared to the contemporary epoch. If we assume that the solar constant in the first half of the Phanerozoic was 4% less than at present and take into account the data on the sensitivity of the thermal regime to heat income variations that was presented in Chapter 3, the corresponding decrease in the mean air temperature will be 5 to 6° at constant albedo.

The actual differences must have been higher than that since the earth's albedo increased due to the development of polar ice cover in the late Phanerozoic and changes in vegetation composition on the continents. However, if we consider these effects, the above differences are altered by the same relatively small value in both cases and our main conclusion based on the calculation remains untouched. It still holds that the greenhouse effect made up for the lower value of the solar constant in the first half of the Phanerozoic, resulting in a warmer climate than at present.

It is often believed that the CO_2 content in Precambrian time was much higher than in the Phanerozoic. The greenhouse effect associated with it was more intensive and could maintain higher temperatures near the earth's surface at a decreased solar constant as well.

The atmosphere of Venus presents an example of an identical effect of CO_2 on the climate. Although Venus is closer to the sun, it absorbs less radiation than the earth due to its high albedo. Since the CO_2 content of the atmosphere of Venus is very high, it enhances the greenhouse effect, leading to a temperature near its surface of up to a few hundred degrees Celsius.

For most of the earth's early history its atmosphere probably contained enough CO_2 to maintain fairly high air temperatures near the earth's surface. In such a case it would not be necessary to assume the presence of ammonia in the ancient atmosphere of the earth in order to explain the paleoclimatic data. Hart (1978) as well as Owen, Cess, and Ramanathan (1979) have recently come to the same conclusion.

It follows from the above considerations that a relatively warm climate for the greater part of the Phanerozoic is accounted for by a considerably higher CO_2 content in the atmosphere than at present. During an epoch of gradually decreasing CO_2 concentration, corresponding to the Cenozoic era, a cooling trend had to develop (Budyko, 1974).

The quantitative study of this problem requires data on the thermal regime of the relevant epoch. This involves certain difficulties since the error in the temperature determination by the method of isotopic analysis of organic sediments or by other known methods is usually at least a few degrees. Such errors are comparable to the variations in the thermal regime extended over millions of years.

To analyze the secular variations in the air temperature during the late Mesozoic and the Tertiary we can use Emiliani's data on paleotemperatures obtained by the method of isotopic analysis (Emiliani, 1966) and Sinitsyn's information based on a complex study of natural conditions in the past (Sinitsyn, 1965, 1967).

Figure 4.1 shows variations in the difference between the mean paleotemperatures and modern temperatures. Curve 1 is constructed according to Emiliani's data defining secular variations in the mean temperature of the oceanic surface layer. Curve 2 is based on Sinitsyn's data.

Since Emiliani's data cover a restricted area of the west Atlantic offshore zone, it is not clear whether it is possible to use them for determin-

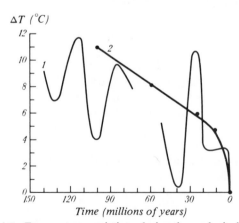

Fig. 4.1 Temperature variations during the geological past.

Table 4.1

Variations in Air Temperature and Carbon Dioxide
Concentration in the Geological Past

	ΔT_{30-80}	ΔT_{0-90}	CO_2 conc. (%)
Contemporary epoch	0	0	0.03
Early Pliocene	9.2°	4.8°	0.06
Early and Middle Miocene	11.8°	6.0°	0.09
Paleocene–Eocene	15.2°	8.2°	0.16
Cretaceous	17.4°	11.0°	0.27

ing global variations in the air temperature. The paleoclimatic research
carried out by Sinitsyn gives more complete information concerning the
thermal regime in the past. Sinitsyn has constructed maps for January and
July temperatures as well as for the annual precipitation totals. These
maps give an idea of the climatic conditions of the extratropical zone (30–
80°N) in the Northern Hemisphere for Europe, Asia, and Africa. They are
based on the climatic relations of vegetation cover and the formation of
sediments. Sinitsyn's set of maps covers the entire Phanerozoic and is
more detailed for the Tertiary.

Utilizing Sinitsyn's data, we can calculate the differences between the
mean annual air temperatures for the periods under consideration and for
the modern epoch for the areas presented in the above maps (ΔT_{30-80}).
Then, extrapolating the variations obtained in the mean latitudinal temperature differences to the equator and the pole, we can find the variations in the mean air temperature differences in the Northern Hemisphere
(ΔT_{0-90}). Table 4.1 gives the results of the calculation for the late Mesozoic and the Tertiary. It also includes CO_2 concentrations based on the
data given in Chapter 2.

We can see that for the period covered by this table the mean air temperature was gradually decreasing, corresponding to a decrease in atmospheric CO_2. In Fig. 4.1 curve 2, which corresponds to the mean temperature change, agrees satisfactorily with curve 1. This proves that the data
obtained by Emiliani are representative.

In Fig. 4.2 the points represent data on the mean air temperature variations near the earth's surface and the different CO_2 concentrations given
in Table 4.1. The dependence of the mean air temperature near the earth's
surface on the CO_2 concentration is also shown in the shape of the curve.
This curve is based on data from Chapter 3 that indicate that with a 0.03–
0.06% variation in CO_2 the mean air temperature rises by 2.5° due to the
increase in the greenhouse effect in the atmosphere and by 2° as a result of
the albedo change due to partial melting of polar ice and changes in vege-

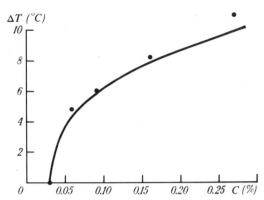

Fig. 4.2   Dependence of mean air temperature on CO_2 concentration.

tation composition. Then it was assumed that the mean temperature increases by 2.5° with each subsequent doubling of the CO_2. Since, for the time interval considered, the effect of solar constant variations on air temperature is an order of magnitude less than that of the CO_2 variations, it was possible to omit it in the computation of the air temperatures.

As can be seen, the curve agrees satisfactorily with the empirical data. It confirms the notion that the major cause of the cooling that developed within the last hundred million years was the CO_2 concentration decrease in the atmosphere.

A great deal of new data on thermal regime variations during the Tertiary, based on paleotemperatures, has been obtained in recent years. Using these data we can get a more detailed notion of the development of the Cenozoic cooling. For instance, the available materials include Buchardt's data on the thermal regime of the surface water of the North Sea for various intervals of the Tertiary (Buchardt, 1978). It follows from this investigation that after the first cooling in the Paleocene, when the sea temperature dropped to 17.5°C, a noticeable warming marked the Eocene (22.5°C). In the Oligocene a sharp cooling occurred (9.5°C), which was followed by a slight warming in the Miocene (11.5°C). Then a new cooling started in the Pliocene (10°C). All these values are average temperatures for the relevant intervals of time.

Comparison of temperatures with CO_2 concentrations in the Tertiary (Fig. 2.1) reveals a striking qualitative agreement between them. This agreement is also evidence for the reliability of the information concerning the evolution of the chemical composition of the atmosphere presented in Chapter 2. It also shows that the concept of atmospheric CO_2 as a determinant of climatic conditions in the Tertiary is quite correct.

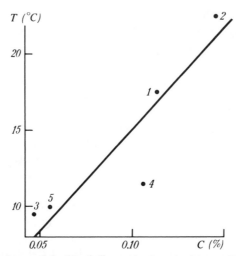

Fig. 4.3 Dependence of the North Sea water temperature on CO_2 concentration.

The quantitative comparison of CO_2 concentrations with data on the thermal regime based on Buchardt's investigation is presented in Fig. 4.3. The logarithm of the CO_2 concentrations is plotted along the x axis. The points designate values referring to the Paleocene (1), Eocene (2), Oligocene (3), Miocene (4), and Pliocene (5). The straight line approaches quite closely the empirical points, which confirms the theoretical conclusions concerning the dependence of the temperature of the lower air layers on the atmospheric CO_2 content. The slope of the line corresponds to a 6.6° temperature increase with a doubling of CO_2 concentration. Considering that the temperature change due to CO_2 variations at the latitude of the North Sea is ~1.5 times larger than the mean global variation (Manabe and Wetherald, 1980), we find that the temperature variations in question correspond to a mean global temperature increase of 4.4°.

This value approaches the previously discussed parameter $\Delta T_c'$. It should be borne in mind that other data on secular temperature variations based on recent paleotemperature measurements for the Tertiary do not always coincide with temperature variations obtained by Buchardt. However, these data usually enable us to reveal the basic regularities derived from Buchardt's work, namely a sharp cooling in the early Oligocene and a slight warming in the Miocene. This coincides with conclusions based on the analysis of atmospheric CO_2 variations.

The conclusion obtained here alters the formerly widespread view that the basic reason for the cooling that created conditions suitable for glaciations was variations in the structure of the earth's surface. As is known, the problem of the effect of the earth's surface structure on climate was first considered by Lyell (1830, 1832, 1833) who compared climatic conditions for the following hypothetical cases: (1) continents are present in high latitudes and (2) oceans are present in the same zone. Lyell held that in these cases the climates must differ greatly. Therefore, he concludes that the evolution of the earth's surface structure was accompanied by considerable climatic fluctuation. Later Ramsey (1910) supposed that the cooling that was favorable for the development of glaciations originated from the upheaval of the continents and the contraction of the ocean surface, which absorbs more solar radiation than the land. Brooks (1949) indicated that a cold climate in high latitudes essentially depended on the presence of polar ice, since ice and snow absorb less solar radiation because of their high albedos. Brooks also emphasized the role of sea currents, which greatly reduce the temperature contrast between low and high latitudes. According to Brooks, the warm and hot climatic conditions at all latitudes that were typical of the most of the earth's history prevailed when the continents were at a low level and widely separated by oceans. A rise in the continental level, especially in the polar regions, led to cooling and the development of glaciations that later expanded to midlatitudes.

Albrecht (1947) considered the upheaval of meridional mountain chains, particularly of the Rocky Mountains and the Andes, to be an important factor in climatic cooling in Quaternary time. According to Albrecht, these mountains make the atmospheric circulation far less intensive, they decrease winter precipitation over the continents in middle and high latitudes, and as a result, they lower condensation heat income.

Some scientists have treated the problem of the effect on climate of variations in the sea bottom relief (Ewing and Donn, 1956, 1958; Rukhin, 1958). They believed that the Quaternary glaciations resulted from the upheaval of the underwater mountain ridge in the North Atlantic that extends from Scotland to Iceland and Greenland. This upheaval suppressed heat transfer by the Gulf Stream, bringing about a cooling of polar latitudes, where cold dry air masses started accumulating. The interaction of these masses with warm damp sea air caused abundant solid precipitation that led to continental glacier expansion.

The question of the role of variations in the relief structure during the Cenozoic cooling can be resolved with the help of numerical climatic models. Some of the results obtained are given in Budyko (1974). They show that a decrease in the meridional heat transfer in the ocean due to a rise in the continental level leads to an air temperature decrease in middle and high latitudes. At the same time, calculations carried out in terms of the semiempirical theory of climate can show that variations in the meridional heat transfer at constant earth albedo have no bearing on either the mean global or hemispheric temperature.

Since the above empirical data indicate that for the last hundred million years the mean air temperature in the Northern Hemisphere decreased considerably, by $\sim 10°$, we can conclude that variations in the relief structure could hardly have caused such a cooling; rather, it was brought about by a decrease in atmospheric CO_2.

This conclusion, however, does not eliminate the possibility that polar glaciations were influenced to a certain extent by the relief structure.

In view of the contemporary position of the Antarctic continent in the high latitudes of the Southern Hemisphere and the restricted interconnections of the Arctic Ocean with the Atlantic and the Pacific, sea currents transfer no heat at all to the southern polar zone and a very small amount to the northern polar regions (Budyko, 1971). This facilitates the development of glaciations in high latitudes, increasing the earth's albedo and decreasing the mean air temperature near the earth's surface. According to the above calculations, however, this decrease does not exceed $2°$ and appears to be far less than that which took place in the Cenozoic.

We can conclude that if a major reason for the Cenozoic cooling was the atmospheric chemical composition change, it is highly probable that the evolution of the earth's surface structure was of importance in originating the polar ice cover at the end of the Cenozoic. This ice cover in the Pleistocene became a center of glaciation advances typical of the ice ages.

Turning to the causes of the advance and retreat of the ice cover in the Quaternary, we should mention that many different hypotheses have been suggested in the past concerning the nature of the Quaternary glaciations. For example, Sarasin and Sarasin (1901) put forward a hypothesis explaining the Quaternary glaciations by a decrease in atmospheric transparency in the epochs of higher volcanic activity. Their hypothesis has been supported by some later investigations (Fucks and Patterson, 1947; Kennet and Thunell, 1975). Arrhenius (1896, 1903) and Chamberlin (1897, 1898, 1899) claimed that the Quaternary glaciations were caused by fluctuations in atmospheric CO_2 concentration.

Many scientists suggested that these glaciations resulted from auto-oscillatory movements in the system that includes the atmosphere, the oceans, and the ice cover. For example, Plass (1956) claimed that the auto-oscillatory nature of changes in the CO_2 concentration in the ocean and the atmosphere was an essential factor in glaciation advance. In his opinion, after an initial small decrease in the atmospheric CO_2 concentration, the temperature near the earth's surface dropped, leading to polar ice expansion. As a result, the volume of oceanic water decreased, causing a rise in the atmospheric CO_2 and, consequently, a new warming. During the warming, the ice began melting, the ocean volume increased, the atmospheric CO_2 decreased, and a new cooling process set in. In a study by Wilson (1964) that was further developed by Flohn (1969) the cause of the Quaternary glaciations was connected with an unstable Antarctic glacier whose margins were subjected to large ice slides, leading to auto-oscillatory processes in the atmosphere. Sergin and Sergin (1969, 1978) hold that the auto-oscillatory processes were determining factors for the Quaternary glaciations. To study this problem they constructed a quantitative model of the atmosphere–ocean–ice-cover system.

A number of works discuss the question of the possibility that this system is almost intransitive (see Chapter 3). If such were the case, the appearance and destruction of the glaciations would have, to a great extent, a random character.

We shall not dwell on the many other hypotheses concerning the origin of the Quaternary glaciations but we mention that the so-called astronomic theory of glaciations attracts very serious attention from modern scientists.

Climate is known to depend on fluctuations in the radiation incident on various latitudinal zones of the earth in different seasons, which are caused by variations in the position of the earth's surface with respect to the sun. This position is dependent on the eccentricity of the earth's orbit, the obliquity of the earth's axis to the ecliptic, and the presession of the equinoxes. All these astronomical elements vary periodically, changing somewhat the amount of radiation that reaches various latitudes in different seasons.

To elucidate the influence of variations in radiation on climate, the author has used the numerical model of the thermal regime for average annual conditions as set forth in Chapter 3 (Budyko, 1968a). The result was that variations in the radiation regime during the last Würm glaciation could cause a southward shift of 1° latitude of the ice cover in the Northern Hemisphere. The actual shift was considerably larger. We should recognize that the determination of mean annual temperatures is not sufficient for estimating the effect of orbital changes on the glaciation, since the glaciation regime depends largely on the thermal conditions of the warm seasons.

Subsequently, Budyko and Vasishcheva (1971) applied another model, also described in Chapter 3, for the study of the climatic conditions of the ice ages, which defined the distribution of the mean latitudinal temperature for various seasons.

As the analysis of the relevant equations can show, calculations of the temperature variations based on either this model or the model for the average annual conditions gave greater accuracy than the mean air temperature computations. This enables the model to be used in the study of climatic change.

The model was used to estimate the average polar ice boundary for the period of time when the radiation income in high latitudes during the warm half-year was noticeably decreased because of astronomical factors. The date on the radiation regime for the indicated periods were derived from the study by Milankovich (1941).

The calculation takes into consideration the dependence of the planetary albedo on the polar ice boundary:

$$\Delta\alpha_{sp} = \frac{Q_{siN}\,\Delta L_{iN}}{Q_{sp}L_0}\,(\alpha_{siN} - \alpha_{sfN}) + \frac{Q_{siS}\,\Delta L_{iS}}{Q_{sp}L_0}\,(\alpha_{siS} - \alpha_{sfS}), \quad (4.1)$$

where $\Delta\alpha_{sp}$ is the planetary albedo change as compared to its modern value, ΔL_i the variation in the area of the latitudinal zone taken up by ice in one of the hemispheres as compared to the present regime, L_0 the area of the globe, α_{si} the albedo of the polar ice zone, α_{sf} the albedo of the ice-free zone where glaciation has spread as a result of variations in the radiation, Q_{si} the radiation at the outer boundary of the atmosphere in the zone with area ΔL_i where the glaciation has spread, and Q_{sp} the mean planetary radiation. The indices N and S designate the Northern and Southern Hemispheres. The calculation neglects the effect of variations in the radiation balance of the ocean surface and in cloudiness compared to the present-day regime.

Some results of the calculations are given in Table 4.2. The table shows that variations in the radiation regime caused by a change in the earth's surface position relative to the sun can result in a significant climatic change. The calculations demonstrate that the mean temperature of the

Table 4.2
Climatic Change in the Ice Ages[a]

Time (thousands of years before 1800 AD)	$\Delta\varphi_N$ (deg)	$\Delta\varphi_S$ (deg)	ΔT (°C)
22.1 (Würm III)	8	5	−5.2
71.9 (Würm II)	10	3	−5.9
116.1 (Würm I)	11	2	−6.5
187.5 (Riss II)	11	0	−6.4
232.4 (Riss I)	12	−4	−7.1

[a] $\Delta\varphi_N$ and $\Delta\varphi_S$ are the decrease in the mean latitude of the polar ice boundary in the Northern and Southern Hemispheres, respectively, compared to their present positions; ΔT is the mean temperature variation during the warm half-year at 65°N.

ice-free zone varies relatively little, by ~ 1°. However, this small variation is accompanied by a noticeable shift of the ice-cover boundary.

The calculations show that the greatest shift of ice boundary in the Northern Hemisphere within the period under consideration was 12° and in the Southern Hemisphere, 5° (the contemporary mean latitudinal ice boundary in the Northern Hemisphere is close to 72°N and in the Southern Hemisphere nearly 63°S).

A considerable temperature drop can be observed in the zone of ice expansion. For instance, at 65°N, advance of ice causes a 5–7° drop in the mean temperature of the warm half-year. We should remember that this applies to sea-level temperatures. The massive continental glaciations are likely to reduce ice-level temperatures even more.

The greatest change calculated for the mean latitudinal ice cover margin in the Northern Hemisphere agrees well with the empirical data. For example, Lamb (1964) found that during the greatest glaciations the mean ice boundary in the Northern Hemisphere reached 57°N, which corresponds to a 15° shift in this boundary as compared with its present position. We have obtained a 12° shift in the same boundary, which approaches Lamb's result.

The same comparison for each of the ice epochs presents difficulties, since data on the mean latitudinal ice cover boundaries do not exist. We can use, however, the available information on the ice cover boundaries in certain areas of the globe. For instance, Zeuner in his monograph (1959) presents data on the areas covered with glaciers during different epochs in central and northern Europe. These areas, which Zeuner expressed as a percentage of the extent of ice cover during the epoch of Mindel II, have been compared to the calculated latitudinal variations in the northern polar ice boundary for the epochs of Riss I, Riss II, Würm I, Würm II, and Würm III as shown in Fig. 4.4. Since Zeuner's magnitudes

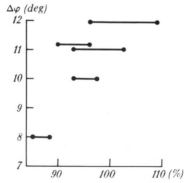

Fig. 4.4 Glacier boundaries during the epochs of glaciations.

of the ice extent for each of these epochs are given as a particular range, the results obtained by comparison are shown here as line segments. The figure shows a relationship between the results in question, which proves that the calculation method can be used for the estimation of comparable parameters of different glaciations.

We emphasize that the above agreement between the results obtained for the polar ice boundary and the paleogeographic data could be achieved on the basis of the feedback relationship between the position of the polar ice and the atmospheric thermal regime. An interesting example of calculations in which the indicated feedback was not considered was presented by Saltzman and Vernekar (1971). They used a numerical model of the mean latitudinal distribution of temperature, wind, evaporation, and precipitation based on the integration of the equations of atmospheric dynamics. With this model they calculated the variations in the temperature distribution near the earth's surface in the Northern Hemisphere that occurred 10 and 25 thousand years ago, as compared to contemporary conditions. The calculation assumed that all the climate-forming factors are constant, except for the radiation at the top of the atmosphere, the distribution of which differed somewhat from the present one because of astronomical factors.

Saltzman and Vernekar found that the greatest temperature difference during the warm half-year did not exceed 1.0° at some latitudes for the indicated periods. They believed that such comparatively small temperature variations could not result in the development of glaciation.

Since Saltzman and Vernekar's estimation did not take into account the feedback between the ice cover and the temperature field, the temperature fluctuations due to radiation change at the top of the atmosphere that they obtained must have been underestimated. It is interesting to compare these results with those of a similar calculation based on our model without considerations of the feedback, i.e., one which considers the albedo of the earth–atmosphere system to be constant while the radiation varies. The results of such a calculation are presented in Fig. 4.5, which shows the distribution of the mean latitudinal temperature differences for the warm half-year for contemporary conditions and for the periods 10 and 25 thousand years ago.

It follows from the figure that the results obtained by different kinds of models without consideration of the feedback between the ice cover and the thermal regime are quite similar. This is noteworthy since the Salzman–Vernekar model differs considerably from our more schematic model. An agreement between the paleoclimatic results obtained with various climatic models reveals a sufficient reliability of these models and shows that semiempirical theories of the thermal regime can be based on many simplified assumptions.

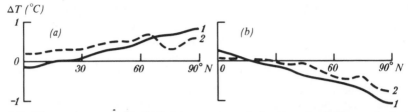

Fig. 4.5 Changes in mean latitudinal air temperatures during the warm half-year, due to the effects of variations in incoming radiation caused by astronomical climate factors: (a) 10^4 years ago; (b) 2.5×10^4 years ago; curve 1, calculated results, by Budyko and Vasishcheva (1971); curve 2, calculated results, by Saltzman and Vernekar (1971).

The agreement between the results of thermal regime variations based on the above semiempirical model of the thermal regime and on more generalized climate theories arises evidently from the confirmation of the basic empirical hypothesis used in the model and expressed by the equation $C = \beta(T - T_p)$ not only for average annual conditions but also for different seasons of the year (see Chapter 3).

It is known that the mean temperature difference between the pole and the equator varies considerably in the course of the year, corresponding to a change in the energy income that thus produces large-scale atmospheric motions. The basic patterns of the global atmospheric circulation, however, do not differ greatly in summer and winter. Therefore the meridional heat transfers during different seasons can be described by similar empirical dependences.

From this point of view, the work by Williams *et al.* (1973) that presents a numerical model of atmospheric circulation during the last Würm glaciation seems quite interesting. Flohn and other scientists held that during the ice ages summer climatic conditions in middle and high latitudes were similar to present winter conditions. Williams *et al.* came to a similar conclusion by physical deduction.

It follows from their study that during the ice ages the summer westward transfer of air masses in middle latitudes of the Northern Hemisphere roughly corresponds to that of contemporary winter conditions. The glaciations resulted in a slight shift of the stable pressure systems and a certain redistribution of precipitation. The atmospheric circulation, however, underwent no abrupt changes during that time. Similar conclusions have been drawn by other scientists (Alyea, 1972; Gates, 1976).

New empirical and theoretical investigations of Quaternary glaciations carried out in recent years have revealed information essential for understanding the physical mechanism of the ice ages.

CLIMAP compiled extensive data on temperatures during the last glaciation, based on isotopic analysis and plankton distribution in various regions of the world ocean. These data have been used to construct world maps of the temperature of surface oceanic waters. An important conclusion implied that during the glaciations the mean temperature of the ice-free ocean surface was only 2° below the present one (CLIMAP Project Members, 1976). This conclusion agrees reasonably well with the above results based on the semiempirical theory of the atmospheric thermal regime.

According to empirical data, the temperature decrease was somewhat greater than the results obtained above. This can be explained by an increase in the earth's surface albedo due to the expansion of arid conditions on the continents during glaciation advance (Cess, 1978).

Hays, Imbrie, and Sheckleton have compared temperature fluctuations for the last 450,000 years (derived from the composition of marine organic sediments in the Southern Hemisphere) with the variations in radiation due to astronomical factors (obtained by calculation) (Hays et al., 1976).

A detailed statistical analysis of temperature variations revealed clearly pronounced fluctuations with periods of 23,000, 42,000, and ~ 100,000 years. The first of these periods coincides with the oscillatory period of the precession of the equinoxes, the second corresponds to the oscillatory period of the inclination of the earth's axis, and the third approaches the oscillatory period of the eccentricity of the earth's orbit. Proceeding from this agreement, the authors concluded that variations in the astronomical factors were the basic reasons for the successive Quaternary glaciations.

The study by Hays et al. attracted great attention and was widely discussed in numerous publications by different experts. Considerable interest was drawn by the main conclusion stated by Imbrie at the Soviet–American Symposium on Paleoclimatology in Moscow in 1976. He said that the long discussion concerning the causes of the Quaternary glaciations was over, since these causes were now known.

The studies by Gates (1976a,b) and Manabe and Hahn (1977) should be mentioned among the new theoretical investigations relating to climatic conditions during the ice ages. These authors used models of the general circulation to study the atmospheric general circulation during the last glaciation. Using the CLIMAP data on temperature distribution over the ocean, the authors were able to elucidate many climatic features of the ice ages. For instance, precipitation was found to decrease noticeably during glaciations, leading to the expansion of insufficient moisture conditions on the continents. This conclusion has been confirmed by an empirical study of Pleistocene natural conditions (Velitchko, 1973).

Some authors applied simplified climatic theories to elucidate causes of the Quaternary glaciations, allowing them to avoid the use of empirical data for certain elements of the past climatic regime. Berger (1973, 1975, 1977a, 1978) carried out many investigations of this kind using a semiempirical model of the atmospheric thermal regime and decided that the astronomical factors had a determining effect on the Quaternary glaciations. Using various parameterized climatic models, Suarez and Held (1976), Pollard (1978), and others have come to similar conclusions. The studies of this trend have emphasized that a realistic description of the Pleistocene glaciations can be made only with the help of climatic models based on the feedback between polar ice and the thermal regime of the atmosphere.

For instance, Schneider and Thompson (1978) have shown that the calculation of temperature variations in high latitudes without consideration of the feedback underestimated the temperature differences in comparison with paleoclimatic data.

Cess and Wronka (1979) have mentioned that for the correct estimation of air temperature variations during glaciations it is necessary to take into account a positive feedback between temperature fluctuations and the variations in the albedo of the continental surface that are determined by the state of the vegetation cover.

Research presenting new and more accurate calculations of variations in those elements of the radiation regime that affected the development of glaciations (Sharaf and Budnikova, 1969; Vernekar, 1972; Berger, 1973, 1976) was also of great significance in studying the origin of the ice ages.

It follows from the paleogeographic data that the polar ice cover appeared millions of years ago, caused by the cooling resulting from a decrease in CO_2. Astronomical factors started to influence the climate significantly only after the development of large polar glaciations, since it is these that made Quaternary climatic conditions so sensitive to minor variations in the climate-forming factors. During earlier epochs, when polar glaciations were relatively small or absent, astronomical factors varied in a manner similar to that in the Quaternary but their climatic effect was insignificant. The extent of this influence seems to correspond to variations in the thermal regime of the atmosphere, calculated on the basis of our model and the Saltzman–Vernekar model, as illustrated by Fig. 4.4. The figure shows that in the absence of polar ice cover fluctuations of the astronomical factors alter the air temperature at all latitudes by no more than 1°. This means that in pre-Quaternary time the astronomical factors did not exert a strong influence on climatic conditions.

The above statements show that at present the causes of climatic cooling during the Cenozoic era, as well as the causes of the Quaternary glaciations, are more or less known.

4.2 PRESENT-DAY CLIMATIC CHANGES

Climate of the Holocene

The post-Pleistocene climatic variations occurred within a comparatively short period of the earth's history. During this time, which is often called the Holocene, we can distinguish several epochs with different climatic conditions.

Many of the works on the paleoclimatology of the Quaternary cited in the preceding subsection present some information on the climate in the Holocene.

The last Würm glaciation reached its greatest extent about 20,000 years before our time. In a few thousand years its area contracted considerably. The climate of the next epoch was comparatively cold and humid in the middle and high latitudes of the Northern Hemisphere. About 12,000 years before our time the climate became much warmer (the Alleröd epoch). However, a cooling trend soon developed. As a result of these climatic fluctuations, the summer air temperature in Europe varied by a few degrees.

Later on the warming trend resumed and 5000 to 7000 years before our time the remains of the continental glaciations disappeared in Europe and North America. It was in that epoch that the postglacial warming reached its peak. Between 5000 and 6000 years before our time the air temperature in the middle latitudes of the Northern Hemisphere is believed to have been about $1-3°$ higher than at present.

At the same time the atmospheric circulation apparently underwent a change. The northward shift of the polar-ice boundary was accompanied by the displacement of a subtropical high-pressure belt to higher latitudes. This resulted in the expansion of arid zones in Europe, Asia, and North America. In addition, precipitation increased in low latitudes over the areas now occupied by deserts. The climate in the Sahara was then comparatively humid, favoring abundant flora and fauna. Still later a tendency toward cooling prevailed, which became especially pronounced at the beginning of the first millennium B.C. The thermal regime change was accompanied by an alteration of the precipitation regime that brought about a gradual approach to its present state.

A significant warming trend took place during the end of the first and the beginning of the second millennia A.D. The polar ice retreated to high latitudes, allowing the Vikings to colonize Greenland and to discover the continent of North America. A subsequent cooling favored a new advance of ice and the Norse colony in Greenland, deprived of any communication with Europe, perished.

The cooling trend that was developing since the 13th century reached its maximum in the early 18th century and was accompanied by the expansion of mountain glaciers. For this reason it is called the Little Ice Age. Then the warming started again and the glaciers retreated. Climate during the second half of the 18th century and the 19th century differed comparatively little from contemporary conditions.

Of great importance in understanding the physical mechanism of contemporary climatic change is the study of climatic variations over the last century, when most of the continental surface was monitored by a network of permanently operating meteorological stations.

The greatest climatic change within the period of instrumental observations started at the end of the 19th century. It was marked by a gradual rise in air temperature at all latitudes in the Northern Hemisphere throughout the year. This warming was especially pronounced at high latitudes and during the cold seasons.

The warming intensified in the 1910s and reached its peak in the 1930s, when the mean air temperature in the Northern Hemisphere was 0.6° higher than at the end of the 19th century. In the 1940s the warming trend changed to a cooling trend, which continued until recently. The cooling progressed slowly and its scale did not approach that of the preceding warming.

Although the evidence of modern climatic change in the Southern Hemisphere is less certain than that for the Northern Hemisphere, there are grounds for believing that the warming also occurred in the Southern Hemisphere in the first half of the 20th century.

In the Northern Hemisphere the air temperature rise was accompanied by a contraction of polar ice, a retreat of the permafrost boundary to higher latitudes, a northward shift of the forest and tundra boundary, and other changes in natural conditions. Of vital importance was a noticeable change in the atmospheric precipitation regime during the warming. In some regions of insufficient moisture conditions, precipitation diminished during the climatic warming, especially in the cold seasons. This resulted in decreased river runoff and a fall in the water level of some land-locked reservoirs. The abrupt drop in the water level of the Caspian Sea that occurred in the 1930s and was caused by the reduced runoff of the Volga River is particularly well known.

In addition the warming brought about more frequent droughts over vast continental areas in mid-latitudes in Europe, Asia, and North America. This climatic change had a definite influence on the national economy of many countries.

Variations in the thermal regime

The general regularities of thermal regime variations in the last century have been studied by Rubinshtein (1946, 1973), Rubinshtein and Polozova (1966), Mitchell (1963), Lamb (1966, 1977), Van Loon and Williams (1976a–c, 1977), and many other scientists. These investigations have established the temporal and spatial inhomogeneity of the above variations.

The works by Rubinshtein and Polozova show, in particular, that mean anomalies of the monthly air temperature at a definite point vary considerably for successive months, the variations often being irregular.

A complex temporal and spatial structure of the temperature anomalies may be influenced to a certain extent by fluctuations in the atmospheric circulation that are random relative to climate variations. Averaging of the temperature anomalies over long periods of time for certain stations or limited regions does not entirely eliminate the effect of this factor, which produces essential difficulties in revealing the regularities of climatic change.

In studying global climatic fluctuations, spatial averaging of the air temperature anomalies acquires particular importance. This approach is based on the following physical considerations.

As described in the previous chapter, the longwave radiation outgoing into outer space has a linear relationship to the air temperature near the earth's surface. That is why the average of the temperature anomalies corresponds to the average of the outgoing radiation anomalies.

The outgoing radiation anomalies for limited regions are to a great extent compensated for when they are averaged in space and their effect on the global climate appears to be limited. The radiation anomalies for vast territories, particularly the anomalies for the entire globe, exert a great influence on climate. The algebraic sum of this anomaly and that of the shortwave radiation that is absorbed by the earth–atmosphere system determines the sign of the mean air temperature variation over the entire planet.

Mitchell (1961, 1963, 1971) and Willett (1974) have calculated the mean anomalies of the air temperature for large areas, based on observational data on the air temperature from many meteorological stations scattered in various geographical regions.

For a more accurate determination of the mean anomalies, we have used maps of the air temperature anomalies and not the observational data of individual stations. These maps present the distribution of the mean monthly temperature anomalies for each month since 1881 in the

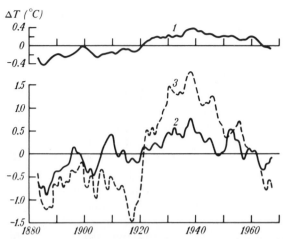

Fig. 4.6 Secular course of air temperature anomalies (5-year running average): curve 1, anomalies of the Northern Hemisphere mean annual temperature; curve 2, anomalies of the 70–85°N latitude belt temperature for the warm half-year; curve 3, the same for the cold half-year.

Northern Hemisphere, the equatorial belt excluded. The latter is not covered well enough by the observational data over the first half of this period for maps of the anomalies to be drawn.

Figure 4.6 shows the secular variation of the air temperature anomalies for the extraequatorial zone of the Northern Hemisphere and for the 70–85°N latitude belt calculated by Spirina (1971) on the basis of the maps. All the data have been averaged over running 5-year periods.

It follows from the figure that a warming trend began at the end of the 19th century in the extraequatorial latitudes of the Northern Hemisphere, with a weakly pronounced maximum just before the turn of the century. This was followed by a slight temperature drop, which was soon replaced by a rapid temperature increase. The warming was especially marked during the cold seasons in the late 1910s and early 1920s. The positive temperature anomaly was greatest at the end of the 1930s; in the 1940s the warming trend was overcome by a cooling trend, which intensified in the 1960s. In the mid-1960s the mean air temperature in the Northern Hemisphere approached the level of the late 1910s.

We believe that the secular temperature variations in the extraequatorial belt of the Northern Hemisphere correspond more or less to the secular temperature variation near the earth's surface over the entire globe. The available global data (which are scarcer than those for the extratropical latitudes of the Northern Hemisphere) show that variations in the mean air temperature also occurred in the equatorial belt and at extratropical latitudes in the Southern Hemisphere. In most zones for which data exist, these variations seem to coincide with those of the zone covered by numerous observational data (Mitchell, 1963).

It has recently been established that, despite numerous peculiarities of thermal regime variations at different points on the globe, there are small areas where thermal regime variations are closely related to the mean air temperature fluctuations in the Northern Hemisphere. For instance, the correlation coefficient between the annual temperature anomalies over the 1899–1976 period for Sverdlovsk and the Northern Hemisphere appears to be greater than 0.5 (Landsberg *et al.*, 1978). Using the data from 9 stations, the authors obtained a correlation coefficient of 0.81 between the temperature fluctuations at the stations and those for the Northern Hemisphere. The result shows that local variations in the thermal regime reflect certain large-scale regularities which are still insufficiently known.

Figure 4.6 shows that secular air temperature variations intensify with latitude, the air temperature being more variable during cold seasons, especially at higher latitudes. We emphasize here that on the whole the thermal regime variations presented in Fig. 4.6 are similar to the results obtained by Mitchell (1963).

In addition to the regular temperature variations over longer periods, the numerous short-term temperature fluctuations shown in Fig. 4.6 are noteworthy. The latter reflect to a great extent the effect of the variability of circulation processes, which is not entirely eliminated by spatial and temporal averaging.

Since the temperature anomalies in low latitudes for individual seasons differed comparatively little, the mean meridional temperature gradient varied less during warm and more during cold seasons.

The secular variations of the mean meridional temperature gradient anomalies are presented in Fig. 4.7 (Budyko and Vinnikov, 1973). The gradient is expressed in degrees of temperature per ten degrees of latitude. The anomalies were determined by the least squares method based on the mean latitudinal temperature for every 5° zone in the interval from 25°N to 70°N and averaged over running 5-year periods.

The figure shows a continuous decrease in the meridional temperature gradient from the 1890s to the 1930s except for two comparatively short periods—the first years of the 20th century and the second half of the 1910s—notable for a temperature gradient rise. In the second half of the 1930s the meridional gradient began rising and in the mid-1960s its anomalies corresponded to those at the turn of the century.

It was generally accepted in the early 1970s that a tendency toward climatic cooling appeared during the last few decades. Since the sign of temperature fluctuations changes relatively rarely, the scientists concerned

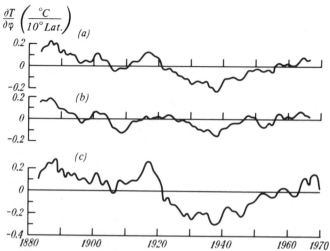

Fig. 4.7 Secular course of anomalies of the meridional temperature gradient over the 25–70°N latitude zone: (a) mean annual anomalies, (b) mean anomalies for the warm half-year, (c) mean anomalies for the cold half-year.

with climatic change almost unanimously believed that the temperature would continue to decrease in the near future.

In this connection, the conjectural further cooling that might result in the development of glaciations and worse climatic conditions for agriculture excited apprehension. The unfavorable climatic conditions due to shortening of the growing season and a decrease in the temperature totals over the vegetation period were expected mostly in countries of the mid-latitudinal belt with surplus moisture conditions. The available data show that in the same latitudinal zone climatic cooling had the opposite effect on agriculture in many regions of continental climate because of an increase in precipitation and a decrease in drought occurrence.

Lamb (1973b) mentioned that more than 20 forecasts of the early 1970s concerning climatic change predicted a cooling trend in the next decades. He indicated, however, a lack of sufficient scientific grounds for these forecasts. Two years later he obtained the first evidence of a possible climatic change toward warming (Lamb et al., 1975). These data describe the thermal regime in the North Atlantic where, in the region of Greenland and the Norwegian seas, the previous tendency toward temperature decrease has changed to warming since the winter of 1970–1971. Willett (1974), analyzing observational data on the air temperature, has come to a similar supposition concerning a new warming trend since the beginning of the 1970s. The data presented by Willett and Lamb called for a detailed study of climatic variations of recent years.

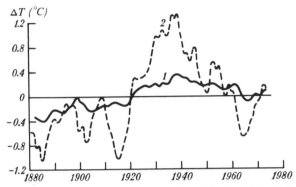

Fig. 4.8 Secular course of air temperature anomalies: (1) for the 17.5–87.5°N latitude zone, (2) for the 72.5–87.5°N latitude zone.

In further studies of the problem (Borzenkova *et al.*, 1976; Budyko and Vinnikov, 1976) the data on variations in the mean air temperature near the earth's surface have been supplemented by new information covering the time up to 1975.

Figure 4.8 shows the secular variations in the air temperature anomalies through the most of the Northern Hemisphere (17.5°N northward) (curve 1). The data are averaged over running 5-year periods.

Analysis of the mean annual temperatures reveals that the trend of the temperature variations in the Northern Hemisphere has recently changed sign. The gradual temperature decrease came to a halt in the mid-1960s and was followed by a temperature rise, which intensified in the late 1960s and early 1970s.

The rate of the temperature rise β over the 1964–1975 period has been estimated as 0.3°/10 yr.

Subsequently Vinnikov and Kovyneva extended the latest observational data on air temperature to 1977 and improved the earlier data used for determining a trend in the temperature variations. They have found $\beta = 0.2°/10$ yr for the 1964–1977 period.

As mentioned above, the air temperature variations have especially large amplitudes in high latitudes and during the cold seasons. Figure 4.8 also shows the data on secular variations in the air temperature in the latitudinal zone north of 72.5°N for average annual conditions (curve 2).

The rate of the mean annual temperature increase for this latitudinal belt is 0.9°/10 yr and 1.3°/10 yr for the temperature of the cold half-year, according to the data for the 1964–1975 period.

The different amplitudes of the low and high latitudinal temperature variations result in significant variations in the mean meridional temperature gradients that accompany the secular air temperature variations of the Northern Hemisphere.

Examination of the secular variations in the mean meridional air temperature gradient within the latitudinal belt 22.5–77.5°N has shown that annual variations in the gradient are 10% of the average magnitude over the period in question. On the average, a 0.1° temperature rise in the Northern Hemisphere due to more intensive warming in high latitudes leads to relative decreases in the meridional gradient of 1% for the cold season and 0.5% for average annual conditions. Since a relationship between anomalies of the mean meridional air temperature gradient and the moisture conditions in the interior continental mid-latitudinal areas has been discovered, the air temperature variations in the Northern Hemisphere affect agriculture, especially in regions of unstable moisture conditions.

The problem of the variations in the mean air temperature for the last years has attracted the attention of a number of scientists. Damon and Kunen (1976) discovered evidence for warming in the Southern Hemisphere and in the Antarctic in particular. However, we should bear in mind that estimation of the mean air temperature and its variations in the Southern Hemisphere is fraught with great difficulties because of relatively few meteorological stations in this area.

Angell and Korshover (1977, 1978) have used observational data from 63 aerological stations to study air temperature variations. In their first paper they concluded that from 1958 to 1965 the mean air temperature near the earth's surface as well as in the troposphere decreased by ~0.3°. After that time the temperature variations were insignificant but the last few years have witnessed a slight warming. Their second paper states that in 1976 the mean temperature was unusually low. However, in the authors' opinion, this does not affect their conclusion concerning the relatively small variations in the temperature after the mid-1960s. Barnett (1978), who used a considerably larger data pool of temperature observations near the earth's surface in the Northern Hermisphere, has obtained results concerning the trend of variations of the average hemispheric temperature that are similar to those of Angell and Korshover.

Painting (1977) has studied fluctuations in the thermal regime that occurred in recent years in middle and high latitudes of the Northern Hemisphere. He has concluded that in the region north of 60°N the previous tendency toward cooling changed to warming in the 1970s. In his opinion, at lower latitudes (from 40 to 60°N) cooling proceeded in the lower troposphere.

A paper by Walsh (1977) presents an analysis of data on the air temperature near the earth's surface in high latitudes of the Northern Hemisphere over the 1954–1975 period. Walsh has discovered a change to warming since the mid-1960s in the region of the previous cooling trend. Considering the data on the mean temperature variations in the 1000–500 mbar layer of the Northern Hemisphere covering the 1943–1976 period, Harley (1978) has revealed cooling and warming periods before and after 1965, respectively, the temperature change being greater in low than in high latitudes. He also mentions a temperature decrease at the very end of the period in question.

A survey of the data on contemporary climatic change (Kukla *et al.,* 1977) presents evidence on variations in the mean air temperature near the earth's surface in the Northern Hemisphere obtained by Yamamoto and co-workers and data on the mean temperature fluctuations in the 500–1000 mbar layer according to Dronia. The first group of data shows in particular that from the 1950s to the mid-1960s a temperature drop occurred in high latitudes of the Northern Hemisphere, which was followed by a temperature rise accompanied by great interannual variability. The data on the air temperature in the lower troposphere show that there was a temperature decrease until the mid-1960s and later relatively small variations in the temperature.

The fact that the above results from different authors do not agree completely is explained by the different data they used in the studies.

First, it is evident that the trends of the mean air temperature variations in the troposphere as a whole and even in its lower layers might not coincide with the trends of temperature variations near the earth's surface. This is not only possible but inevitable if an increase in atmospheric CO_2 affects the temperature variations. (In this case the warming near the earth's surface must be accompanied by cooling in the upper atmospheric layers while the mean temperature of the atmosphere remains stable; see Chapter 1.)

When we are evaluating the reliability of conclusions concerning temperature trends, the data pool and its quality are of great importance. We shall not dwell on this question but emphasize the necessity of having observational data from hundreds of stations that cover more or less evenly both land and ocean for reliable estimation of trends in mean air temperature variations. Unfortunately, not all investigations meet this requirement.

Some authors commit the obvious mistake evaluating air temperature trends on the basis of short-term observations, sometimes carried out for only a year. For instance, in Harley (1978), a conclusion based on the abnormally low 1976 temperature became the grounds for the supposition of the beginning of a new cooling period.

Though it is quite clear that such an approach is wrong when we are defining temperature trends, we mention here that the mean air temperature in the Northern Hemisphere was higher than normal in 1977, and that is why the mean air temperature over two years—1976 and 1977—differed little from the previous mean value.

It is worth noting the conclusion that in the middle or late 1960s global cooling ceased or changed to a warming trend (the latter is more probable) that was most pronounced in high latitudes of the Northern Hemisphere. This conclusion can be drawn even from investigations that are based on not entirely homogeneous materials.

The study of modern changes in the thermal regime of the atmosphere draws attention to the data on the temperature fluctuations of oceanic waters. These data have been analyzed by Grigor'eva and Strokina (1977), who constructed diagrams of the secular variations in the anomalies of water temperature in the Barents Sea and in the Atlantic from 30 to 40°N and to 60 to 70°W (Fig. 4.9). If we compare this figure with Fig. 4.8, we shall see that water temperature variations in these regions agree well with the fluctuations in the mean air temperature, with a small phase lag that is evidently explained by the influence of the thermal inertia of oceanic waters. The paper mentions, however, that the secular course of water temperature in other ocean areas is rather diverse and in some cases does not coincide with variations in the mean global air temperature. The 1951–1960 observational data from weather ships show in particular that in the western and southern areas of the North Atlantic the secular variations of the water temperature were opposite to those in the eastern and northern areas. A similar conclusion has been drawn by Teich (1971). The warming tendency based on recent data for the areas covered by Fig. 4.9 does not reveal itself in the observations of water temperature carried out in the North Pacific (Kukla *et al.*, 1977).

The temperature of the ocean surface is believed to depend on global variations in the thermal regime, not directly, but mainly through fluctua-

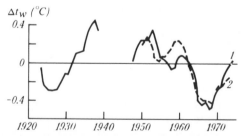

Fig. 4.9 Secular course of anomalies of the mean annual temperature: (1) in the 0–200-m water layer of the Barentz Sea (70.5–72.5°N, 33.5°E) and (2) on the surface of the North Atlantic (30–40°N, 60–70°W).

tions in the system of oceanic currents. For this reason, it is difficult to use data for individual areas for evaluating the thermal state of the world ocean. This, in turn, makes it more difficult to study the relationships between air temperature fluctuations and variations in oceanic temperature, since there is no data on the secular variations of the mean water temperature for the entire surface of the ocean.

We can draw some essential conclusions concerning fluctuations in the thermal regime on the basis of observations of the state of polar sea ice. It should be borne in mind that sea ice boundaries depend strongly on air temperature (Budyko, 1971), and, at the same time, ice position greatly affects the thermal regime of the atmosphere. At high latitudes the air temperature over the ice-free ocean surface during the cold seasons drops only to a few degrees below zero, since under such conditions the ocean releases plenty of heat into the atmosphere. Under the same conditions, in the presence of ice cover that restricts heat flow from the ocean into the atmosphere to a great extent, the temperature of the lower air layer can fall to 10 to 20° or more below zero.

Consequently, the migration of the average boundary of polar sea ice greatly influences the air temperature in relevant areas, especially in the cold seasons. Unfortunately, observational data on the polar sea ice boundary for the period of modern climatic change are rather limited. For the entire last century, data on the ice position are available mainly for the Atlantic sector of the Arctic and the adjacent seas. For the other areas of the Arctic Ocean, as well as for the Antarctic, data on the polar sea ice boundary refer only to the last 20 to 30 years.

Attempts have already been made to elucidate the relationship between climatic change and migration of ice cover in the North Atlantic, in the Barents and Kara Seas.

Rubinshtein and Polozova (1966) have mentioned that since the 1920s the ice cover of the seas in the Atlantic sector of the Arctic has been decreasing. That process continued after the warming trend stopped and it was going on in the Barents Sea until the mid-1950s. Only after that time did the ice gradually start to increase. As for the offshore ice of Iceland, it decreased from the end of the last century until the 1940s, when it began to increase.

In describing changes in the ice boundary in the Arctic, the result obtained by Flohn (SMIC, 1971) is noteworthy. He calculated the differences between the mean latitudinal temperature averages over the year in the Northern Hemisphere for the periods 1931–1960 and 1961–1970. Flohn's diagram shows that from the equator to 50°N this difference approaches zero; in polar latitudes it rises sharply, reaching its maximum of ~ 1.0°C between 70°N and 80°N and then decreasing. This distribution is

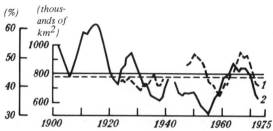

Fig. 4.10 Secular course of the extent of ice (1) in the Greenland Sea (thousands of square kilometers) and (2) in the Barentz Sea (percentage of area).

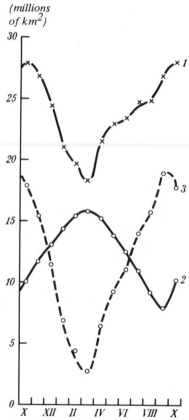

Fig. 4.11 Seasonal variations in the extent of sea ice (1) in the world ocean, (2) in the Northern Hemisphere, and (3) in the Southern Hemisphere.

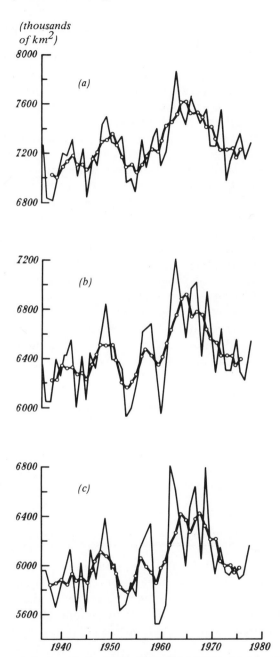

Fig. 4.12 Changes in ice area in the Arctic Ocean for (a) July, (b) August, and (c) September; (○—○) smoothed values, (——) data for every year.

possibly connected with changes in the mean polar ice boundary, which migrated to lower latitudes.

Figure 4.10 presents the secular variations in the ice of the Greenland Sea (curve 1) and the Barents Sea (curve 2) from observational data. Comparison of Figs. 4.8 and 4.10 shows that in this case variations in the amount of ice agree well with fluctuations in the mean air temperature in the Northern Hemisphere. In particular, the ice area decreases rapidly with the development of the warming trend that began in the mid-1960s.

Among other investigations of Arctic Sea ice, we mention here the work of Sanderson (1975), which called attention to a decrease in the amount of ice in most areas of the Arctic in the 1969–1974 period. A detailed description of modern changes in the ice cover of the Arctic Sea is given in a paper by Zakharov and Strokina (1978). We present here a diagram from this paper (see Fig. 4.11) showing seasonal variations in the sea ice area in the world ocean (curve 1), in the Northern Hemisphere (curve 2), and in the Southern Hemisphere (curve 3).

Figure 4.12 illustrates variations in the ice area in the Arctic Sea over the last decades for three months (July, August, and September) (also from Zakharov and Strokina, 1978). The figure shows that from the beginning of the 1940s to the mid-1960s the ice area increased by more than 10%, and since the mid-1960s it decreased by ∼10%. Such a change in the ice cover is in complete conformity with the fluctuations in the mean air temperature at high latitudes.

There are data showing that in the first half of the 1970s the sea ice area also decreased in the Southern Hemisphere (Kukla *et al.*, 1977).

Changes in precipitation

Drozdov, Lamb, and others have established that, during the epochs of warming and cooling, precipitation in various regions of the earth varied to a great extent.

Figure 4.13 presents secular variations in the total precipitation for the cold seasons of the year (from November to March) in the steppe and forest-steppe zone of the USSR. The precipitation totals are calculated for running 5-year periods according to observations carried out at 21 stations. Comparing Figs. 4.8 and 4.13 we can see that a rise in the mean air temperature and a decrease in the meridional temperature gradient are accompanied by a decrease in precipitation in regions of unstable moisture conditions.

Drozdov and Grigor'eva (1963, 1971) have mentioned that during the epoch of the greatest warming (for instance, 1930s) droughts covering

Fig. 4.13   Secular course of total precipitation during the cold period in steppe and forest-steppe zones in the USSR.

vast expanses of insufficient moisture zones of the USSR and North America were more frequent than in the previous and subsequent decades. As is known, in this very period a drastic drop (about 170 cm) in the level of the Caspian Sea took place, caused by decreased precipitation in the Volga River basin.

Shnitnikov (1969) and other investigators have noted that fluctuations in the level of the Caspian Sea are similar in many respects to variations in the levels of several European lakes and consequently they reflect large-scale anomalies in the regime of atmospheric precipitation.

It is possible that the secular variations in the precipitation are to a great extent dependent on variations in the meridional temperature gradient, which affects the nature of the atmospheric circulation.

The intensity of water vapor transfer from the ocean to inland regions varies with changes in the meridional temperature gradient. A decrease in the meridional gradient is accompanied by weakening of the water vapor flow from the ocean to the interior of the mid-latitudinal belt of the continents and by a drop in the amount of precipitation over most of the inland areas. The opposite occurs when the meridional temperature gradient increases.

This conclusion is well confirmed by the observational data. For example, Fig. 4.14 shows the dependence of the total precipitation during the cold period of the year in the steppe and forest-steppe zone of the USSR on the meridional temperature gradient over the running 5-year periods from 1891 to 1960. It follows from the figure that there is a relationship between the parameters in question, which is described by a correlation coefficient of 0.78. As can be seen, the total precipitation over this territory varied by almost 50% from its mean value in some 5-year periods. Such precipitation changes over vast areas obviously greatly affect river runoff and crop yields, which are strongly dependent on precipitation in regions of insufficient moisture conditions. Variations in the level of the Caspian Sea, which are closely connected with the precipitation regime in

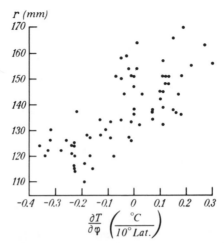

Fig. 4.14 Dependence of precipitation for the cold period in steppe and forest-steppe areas in the USSR on anomalies of meridional temperature gradient in latitudes 25–70°N.

the Volga River basin, clearly reveal the effect of precipitation fluctuations on river runnoff.

It is interesting to compare changes in moisture conditions in the middle latitudes on the continents in the northern part of the Eastern and Western Hemispheres. To study the interconnections between these variations, deviations of the wheat crop from smoothed secular means in the USA over 5-year periods (1910–1971) have been used. These data are the basis for the conclusion that in some 5-year periods the wheat yield on the whole in the USA varied by up to 16% of the normal yields. These variations in the harvest are explained to a great extent by fluctuations in the moisture conditions, which follow, in particular, the coincidence of noticeably decreased harvests with more frequent drought occurrence in the period of warming. Since there are no similar data for vast territories in Eurasia, we used data concerning fluctuations in the Caspian Sea level to reflect moisture conditions.

In Fig. 4.15 the wheat yield anomalies in the USA are compared to variations in the Caspian Sea level ΔH during the same periods. A fairly close relationship has been discovered between these parameters, which is described by a correlation coefficient of 0.74. This relationship confirms that the abovementioned mechanism of the precipitation change has identical effects on the regime of moisture conditions in middle latitudes on different continents.

This conclusion is also confirmed by direct observational data on the precipitation regime in North America. Landsberg (1960) compared air

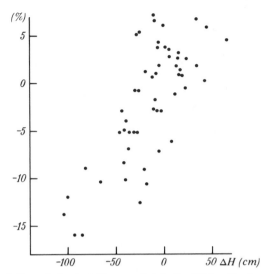

Fig. 4.15 A correlation of wheat yield anomalies in the USA with changes in the Caspian Sea level.

temperature and atmospheric precipitation totals in the USA over two 25-year periods: 1906–1930 and 1931–1955. During the second period the mean annual air temperature near the earth's surface was 0.5° higher than during the first.

The comparison of the amount of precipitation over these periods is not ideal for the purpose of determining the effect of the meridional temperature gradients on the precipitation regime because an abrupt change in the meridional gradient took place, not at the time between the given periods, but around 1920. Nevertheless, Fig. 4.8 clearly shows that during the second period considered by Landsberg the mean meridional temperature gradient was less than during the first period. According to Landsberg, annual precipitation totals over most of the USA territory were smaller during the second period than during the first, which agrees with the above conclusion.

A detailed study of the effect of variations in the meridional temperature gradient on the precipitation regime was carried out by Drozdov and Grigor'eva (1963, 1971). They have found that in high latitudes the general characteristics of precipitation variations during warming or cooling periods are rather complicated, but in areas of insufficient moisture conditions in mid-latitudes, a tendency prevails of precipitation increase with a temperature decrease in the Arctic. Drozdov and Grigor'eva attributed this effect to a more intensive water vapor transfer into the interior of the

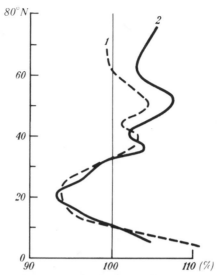

Fig. 4.16 Latitude distribution of total precipitation: (1) from data by Lamb (1974); (2) from data by Borzenkova *et al.* (1976).

continents when there is an increased temperature contrast between high and low latitudes.

   This concept agrees well with the results obtained by Lamb (1974), who constructed world maps of precipitation anomalies for the periods of increased and decreased mean air temperatures near the earth's surface. It can be seen from these maps that during global cooling the total precipitation increased over most of the continental surface in middle latitudes, decreased in the subtropical and tropical zones of high pressure belts, and increased in the equatorial latitudes. Lamb's conclusions have been corroborated by Borzenkova *et al.* (1976), who studied the relationship of the precipitation anomalies with the mean air temperature anomalies in some important agricultural regions of the world. Figure 4.16 presents the relative anomalies of the mean latitudinal precipitation on the surface of the continents during a global cooling period. Curve 1 is constructed from Lamb's data for the entire land surface and curve 2 is based on Borzenkova's data for 9 regions on different continents. The agreement of the curves shows that there is a certain relationship between the distribution of atmospheric precipitation and global variations in the mean air temperature.

   Of great interest is the problem of the relationship of the occurrence of drastic droughts and global climatic fluctuations. Rauner (1979) has noticed that droughts in the USSR and in the USA often occur at the same time. This tendency shows the effect of global factors on drought occurrence. Rauner also presents data on the frequency of prolonged and severe droughts embracing most of the crop regions of the Soviet Union from 1815 to 1976. For our purpose we divide this interval into two periods: 1815–1919 and 1920–1976. The data on the thermal regime show that the mean air temperature for the Northern Hemisphere, averaged over 5-year intervals, was, as a rule, lower during the first period than during the second for any of the 5-year intervals. According to Rauner, the frequency of severe droughts was 1.1/10 yr during the first period and 1.9/10 yr during the second. This conclusion agrees well with the results of other investigations given above.

Causes of climatic change

Reasons for the present change in climate have been found comparatively recently as a result of research accomplished during the past few decades.

Humphreys (1913, 1929) was the first to express views regarding the relationship between contemporary climatic change and volcanic activity. His works had already established that the average amount of direct solar radiation incident on the earth's surface under cloudless conditions varies noticeably for different years. These variations can be seen quite clearly from plots of the secular variations of the direct radiation based on observational data from actinometric stations.

The same curves show that these yearly changes in the direct radiation also vary, on the average over longer periods such as decades.

It is interesting to compare secular variations of the temperature of the Northern Hemisphere to those of radiation incident on the earth's surface. With this aim in mind, an analysis of the actinometric observations over the 1880–1965 period for some European and American stations in the 40–60°N latitudinal belt that had the most long-term data series has been carried out, and a curve of the secular variations in the direct radiation under clear sky conditions averaged over all these stations has been constructed (Budyko and Pivovarova, 1967; Pivovarova, 1968). Figure 4.17 presents solar radiation smoothed over running 10-year periods for the indicated time intervals (curve b). As can be seen, the solar radiation had two maxima, the first being short-term at the end of the 19th century and the second, a longer one, with the greatest values for radiation in the 1930s.

We can state two suppositions concerning the causes of variations in direct radiation under clear sky conditions. The first deals with a relation-

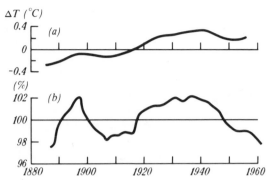

Fig. 4.17   Secular course of anomalies of (a) temperature and (b) direct radiation.

ship of these variations to fluctuations in the astronomical solar constant (the luminosity of the sun) and the second with fluctuations in the so-called meteorological solar constant, i.e., the amount of radiation incident on the upper boundary of the troposphere, which can vary at constant solar luminosity because of the unstable transparency of the stratosphere. The first hypothesis was proposed in several works, for example, the study carried out by Bossolasco *et al.* (1964). In this work the data of three actinometric stations laid the basis for the conclusion that the solar constant increases with the intensification of solar activity (described by Wolf numbers) up to a certain limit, and with a further increase in solar activity, the solar constant decreases.

This concept of fluctuations in the astronomical solar constant was not supported by most specialists in the field of actinometry, who believed that the constant did not vary within the accuracy of measurements (Allen, 1958; Ångström, 1968). To clear up the physical nature of the variations in the solar radiation revealed through the data presented in Figure 4.17, radiation anomalies have been calculated for various altitudes of the sun. Obviously, if relative variations in direct solar radiation are connected with fluctuations in the astronomical solar constant, they cannot depend on the altitude of the sun, whereas solar radiation changes due to the unstable transparency of the stratosphere must differ considerably with different altitudes of the sun. An analysis of the available data has shown that the latter dependence does exist and agrees quite well quantitatively with the results of calculations of the atmospheric transparency effect on the radiation based on formulas of atmospheric optics. As a result, a conclusion has been drawn concerning the presence of fluctuations in the meteorological solar constant while the astronomical solar constant is practically invariable. The indicated conclusion is important in understanding the difficulty of revealing relationships between solar activity and climatic change (Khromov, 1973).

To elucidate the mechanism of modern climatic change we compare curve b (Fig. 4.17) with the curve of secular temperature variations smoothed over running 10-year periods (curve a). It is evident that curves a and b are very much alike. For instance, there are two maxima in each curve, one of them related to the end of the nineteenth century and the other (the major one) to the 1930s. At the same time there are certain differences between the curves. For example, the first maximum is more pronounced for the secular variations of radiation than for the secular temperature change.

Since curves a and b are similar, it is possible to believe that radiation fluctuations due to unstable atmospheric transparency are essential for climatic change. To clear up this problem we should calculate temperature variations as a function of atmospheric transparency fluctuations for shortwave radiation.

The above-mentioned studies by Humphreys show that global fluctuations in atmospheric transparency are primarily caused by relatively small aerosol particles that stay for a long time in the lower stratosphere. Humphreys (1913, 1929) believed that the smallest of these particles can remain in the atmosphere for a few years. These particles have an insignificant effect on longwave radiation but add considerably to the scattering of shortwave radiation. As a result, the planetary albedo of the earth increases while the radiation absorbed by the earth as a whole decreases.

It should be emphasized that since radiation is scattered by particles mostly in the direction of the incident rays (Mie effect), the direct radiation decreases more than the total solar radiation as a result of scattering. Since the thermal regime of the earth is affected by the total radiation variations, to estimate the stratospheric aerosol effect on climate we should determine the total radiation response to stratospheric aerosol particles. For this purpose we can apply the method of calculation used by Shifrin and his colleagues in investigating atmospheric optics (Shifrin and Minin, 1957; Shifrin and Pyatovskaya, 1959).

The results of the investigations were used for determining the ratio of the total radiation decrease to the direct radiation decrease for average conditions at different latitudes in the presence of stratospheric aerosol. The results are given in Table 4.3.

According to Shifrin's calculations, the values in Table 4.3 are almost independent of the particle size if the prevailing diameter varies from 0.02–0.03 to 0.2–0.3 μm.

In evaluating the effect of fluctuations of the direct radiation on the mean air temperature near the earth's surface it is necessary to consider

Table 4.3
Effects of Stratospheric Aerosol on Radiation Regime

Latitude (deg)	90	80	70	60	50	40	30	20	10	0
Total to direct radiation decrease ratio	0.24	0.23	0.22	0.21	0.19	0.18	0.16	0.14	0.13	0.13

the dependence of the mean temperature on the incoming solar radiation. Calculations of the mean temperature variations near the ground with varying radiation income show that a 1% change in radiation leads to 1.1–1.5° fluctuations in the mean temperature near the earth's surface at a constant albedo of the earth–atmosphere system (see Chapter 3).

Let us compare the radiation and thermal regime of the earth for two 30-year periods: 1888–1917 and 1918–1947. As seen in Fig. 4.17, the direct radiation was 2.0% higher during the second period than during the first. According to Table 4.3, the average weighted ratio of total radiation change to direct radiation variations for the hemisphere is 0.16. Taking this into account, we can find that the total radiation was 0.3% higher during the second period. Such an increase in the total radiation corresponds to a 0.33–0.45° mean temperature rise. During these periods the actual temperature difference based on the data in Fig. 4.6 is 0.33°, which agrees reasonably well with the results of calculations (Budyko, 1969a).

For a more detailed study of changes in the radiation and their effect on the air temperature, we shall use the model of the atmospheric thermal regime for different seasons that was presented in Chapter 3.

Figure 4.18 shows a section of the curve (Fig. 4.17, curve b) of secular variations in the smoothed anomalies of direct radiation incident on the earth's surface under clear sky conditions for the 1910–1950 period. The diagram is based on data from several actinometric stations in Europe and

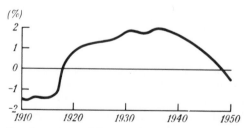

Fig. 4.18   Secular course of direct radiation anomalies for 1910–1950.

North America in the latitudinal zone of 40–60°N. Relying on the above hypothesis, we shall assume that the direct radiation changes are mainly accounted for by fluctuations in the transparency of the lower stratosphere, caused by variations in the concentration of aerosol particles.

We can determine the secular variations in the direct radiation at different latitudes both for average annual conditions and for individual seasons. For this purpose we use the data presented in Fig. 4.18, taking into account the relative optical depth of the aerosol layer and assuming that the stratospheric aerosol content varies little at different latitudes of the Northern Hemisphere.

As has been already mentioned, the total radiation changes due to fluctuations in aerosol concentration represent a small portion of the variations in the direct radiation. We can calculate the temporal variations in the total radiation at different latitudinal belts on the basis of the above data, taking account of the dependence of the total to direct radiation variations ratio on the altitude of the sun.

Figure 4.19 shows the ratio of the total radiation changes at different latitudes of the Northern Hemisphere to the direct radiation variations at 50°N for the warm and cold seasons of the year.

To calculate temperature variations corresponding to the above data on radiation changes, the relationship of the thermal regime and the ice cover should be taken into account. It is evident that in the calculation we cannot use the assumption that the state of the ocean–polar-ice–atmosphere system is stationary. This supposition can be true only for long periods of time because the thermal inertia of the ocean and especially the continental ice cover is large. Therefore, such an assumption cannot always be used in the study of modern climatic changes, which have been at work only for a few decades.

At the same time, we cannot ignore variations in the polar ice regime that are connected with climatic change. For example, according to observational data, warming in the Arctic in the 1930s led to a contraction of the sea ice area by approximately 10% (Ahlman, 1953; Chizhov and Tareeva, 1969).

An accurate estimation of nonstationary processes in the ocean–polar-ice–atmosphere system presents certain difficulties, especially since our knowledge of the heat exchange mechanism between the surface and deeper layers of the ocean waters is inadequate. Therefore, we think it expedient to simplify numerical modeling of present climatic change and roughly assume a certain relationship between the polar ice area and exterior climate-forming factors for the indicated time periods. After the parameters of this relationship are determined from empirical data, it can serve as a substitute equation replacing the expression that relates the

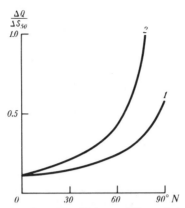

Fig. 4.19 Relation of variations in total radiation at different latitudes in the Northern Hemisphere to variations in direct radiation at 50°N: curve 1, the warm season; curve 2, the cold season.

polar ice area to the thermal regime elements under stationary conditions in our model of the thermal regime.

In view of the semiempirical theory of the thermal regime, the relative variations in the polar ice area are roughly proportional to radiation fluctuations. This dependence is valid for comparatively small variations in the polar ice area and can be written in the form

$$-\Delta P/P = \mu \, \Delta Q_\mathrm{p}/Q_\mathrm{p}, \qquad (4.2)$$

where $\Delta P/P$ is the relative change in the polar ice area, $\Delta Q_\mathrm{p}/Q_\mathrm{p}$ the relative change in the total planetary radiation, and μ a dimensionless coefficient.

For contemporary climate changes, μ can be determined from the empirical data on total radiation and ice cover fluctuations during the Arctic warming. It can be seen from Figs. 4.16 and 4.17 that at that time there was an $\sim 2\%$ increase in direct radiation in the region covered by actinometric stations, i.e., at 50°N, compared to the previous period.

Taking into account this estimate as well as the direct to total radiation ratio and an assumed 10% decrease in the polar ice area during the Arctic warming, we find $\mu = 40$. This is less than a similar coefficient based on the semiempirical theory of the thermal regime under stationary conditions.

It should be noted that this coefficient is a function of the time period covered by the data used for its calculation and the accuracy of calculation is not very high in this case. However, numerical experiments have shown that the error in the determination of the coefficient μ affects the

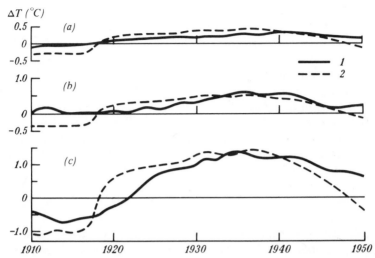

Fig. 4.20   Secular course of air temperature anomalies: (a) Northern Hemisphere; (b) 70–80°N for the warm half-year; (c) 70–80°N for the cold half-year; (———) observed data; (– –) calculated results.

results of calculations of the air temperature distribution comparatively little.

We can calculate the temperature variations at different latitudes for the period in question on the basis of the data from Fig. 4.18 and the numerical model of the thermal regime comprising the Eq. (4.2) instead of the assumed relation between the ice boundary and the air temperature. The results of this calculation are shown in Fig. 4.20 by the dashed curves, which seem to be quite close to the observed temperature variations smoothed over 10-year periods (solid curves).

Attention should be drawn to the fact that the calculated temperature variations would occur somewhat earlier than those based on observational data. The lag of the observed temperature variations is evidently explained by the influence of the inertia of the ocean–polar-ice–atmosphere system. Good agreement of the calculated secular variations in the air temperature with the observational data is achieved with the help of the numerical model of the thermal regime, whose empirical parameters are determined without the data on temperature variations and which consequently yields results independent of the empirical data used in the comparison.

Examination of calculated temperature variations at different latitudes shows that for most of the Northern Hemisphere the temperature rise in the 1920–1930s was caused by increased total radiation coming to the earth's surface. The temperature rise during the warm half-year as determined by calculation and as observed differs little at different latitudes. As can be seen from the calculated values, a screening effect by the aerosol in high latitudes is intensified by the growth of its optical mass and an increase in the total to direct radiation ratio with a decrease in the average altitude of the sun. During the warm seasons the radiation effect on the thermal regime in high latitudes becomes less intensive due to a decrease in the solar radiation with increasing latitude. The above factors act in opposite directions and as a result temperature variations in high latitudes are only slightly increased compared to those in low latitudes.

In the cold seasons the temperature variations in low and middle latitudes differ little from the corresponding fluctuations during the warm seasons. However, they increase sharply in high latitudes (mainly in the 70–80°N latitudinal belt).

Calculations show that in this case temperature variations were only slightly associated with fluctuations in the solar radiation in the same season because during the cold half-year the radiation in high latitudes is very small and does not significantly affect the thermal regime of the atmosphere. The major cause of the temperature change in this case was the change in polar sea ice area, which added appreciably to the air temperature rise in the cold seasons. The effect of change in polar ice position on the temperature of the warm half-year was relatively small.

Savinov (1913), Kimball (1918), Kalitin (1920), and others have established that after the great explosive volcanic eruptions an abrupt decrease in the solar radiation took place.

In such cases the direct radiation averaged over vast territories may be reduced by 10–20% over several months or years. Figure 4.21 presents an example of such a radiation reduction. It shows the change in the ratio between the monthly means of the direct radiation under a clear sky and its normal values after the eruptions of the volcano Katmai in Alaska. The curve is constructed from the observational data of several actinometric stations in Europe and America. It shows that in certain months the atmospheric aerosol reduced the direct radiation by more than 20%.

In some areas the direct radiation decrease was even more significant. For example, in Pavlovsk (in the vicinity of Petersburg), which is situated at a great distance from Alaska, the solar radiation was 35% below the normal value for half a year. Similar radiation changes also occurred after the volcano Krakatoa (in Indonesia) erupted in 1883. In both cases, after the volcanic eruptions, vast territories witnessed abnormal optical phenomena in the atmosphere, which confirmed the planetary nature of the changes in the radiation regime caused by the distribution of stratospheric aerosol.

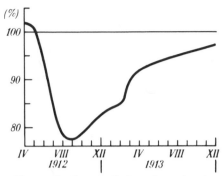

Fig. 4.21   Changes in direct radiation after volcanic eruptions.

A better opportunity to study the effect of volcanic eruptions on the solar radiation regime appeared with the establishment of a world network of actinometric stations that was formed mainly in the 1950s as a result of preparations for the International Geophysical Year. Since then, the first large volcanic eruption of an explosive nature occurred in Indonesia in March 1963 (the volcano Agung on the island of Bali). The effect of this eruption on the radiation regime has been studied with far greater accuracy than was possible for the previous eruptions.

Shortly after the Agung eruption, the effect on the radiation income was detected in various regions of the world (Burdecki, 1964; Flowers and Viebrock, 1965; Dyer and Hicks, 1965, 1968; Budyko and Pivovarova, 1967). To estimate the effect of the Agung eruption on the radiation regime of the USSR, Budyko and Pivovarova have treated observational data on the intensity of the direct radiation at noon from a number of actinometric stations in the Soviet Union over the 1957–1966 period.

They established that from 1957 to November 1963 the monthly values of the direct radiation averaged over 22 stations situated between 40 and 60°N varied comparatively little. Since December 1963, the values decreased abruptly (see Table 4.4). A systematic change was registered since the end of 1963, although certain fluctuations in the intensity of direct radiation from month to month, apparently due to instability of the atmospheric circulation, have been observed earlier.

Since the intensity of direct radiation decreases much more in winter than in summer, it is quite evident that the reduction is explained by fluctuations in atmospheric transparency and not by solar constant fluctuations.

To compare variations in the radiation over USSR territory with those over other regions of the globe on the basis of observations carried out at

Table 4.4
Deviation of the Direct Radiation Intensity from the Normal over the USSR (in %)

	Month												
Year	I	II	III	IV	V	VI	VII	VIII	IX	X	XI	XII	Annual
1963	−1	3	−2	2	2	2	2	1	0	−1	−2	−10	0
1964	−10	−3	−6	−7	−2	−1	−3	−2	−3	−6	−8	−16	−5
1965	−15	−7	−11	−5	−4	−3	−3	−2	−5	−2	−9	−12	−6
1966	−10	−9	−7	−4	−2	−1	1	0	−2	−4	−7	−7	−4

several stations outside the Soviet Union, differences in the intensity of direct radiation at noon in 1964 and 1958 have been calculated. The transparency of the atmosphere in 1958 was assumed to approach the average conditions of the late 1950s and early 1960s. The results of our calculation are presented in Table 4.5.

The data from the table clearly show that the decrease in the direct radiation that occurred in 1964 was also observed in Western Europe, North America, and the central Pacific. All the stations whose data have been included in the table are situated in relatively low latitudes where the average altitude of the sun varied little through the year. Therefore, in these cases there is no appreciable annual variation of the differences in the direct radiation intensity.

The observational data on scattered radiation have shown that since 1963 a drop in the direct radiation was accompanied by a sharp rise in the scattered radiation throughout USSR territory. This follows, for example, from Table 4.6, which shows relative variations in the scattered radiation at noon, under clear sky, averaged for 22 Soviet stations.

If we compare the absolute values of the decrease in the direct radiation incident on the horizontal surface and the increase in the scattered radiation over the USSR territory, it can be established that the total radiation varied comparatively little over the period considered. Although it is difficult to determine the value of this change from the empirical data, it can be calculated, as shown above, on the basis of the direct radiation variations.

Dyer and Hicks (1968) have studied the effect of the Agung eruption on the direct radiation using observations made at several dozen actinometric stations covering both the Northern and the Southern Hemispheres. They have found that the aerosol particles that decreased radiation spread over the entire globe, from the South Pole to the high latitudes of the Northern Hemisphere. It took only a few months for them to diffuse and the high concentration that considerably reduced radiation

Table 4.5

Difference in the Intensity of Direct Radiation in 1964 and 1958 (in %)

Station	Month												Year
	I	II	III	IV	V	VI	VII	VIII	IX	X	XI	XII	
Porto, Lisbon, and Faro (Portugal)	−7	−10	−6	−11	−9	−9	−10	−7	−12	−10	−12	−14	−10
Blue Hill and Albuquerque (North America)	−12	−5	−5	−3	−4	−4	−4	−4	−6	−5	−9	−4	−6
Mauna-Loa (Hawaii)	—	−4	—	−7	−6	—	−6	−4	−6	−6	−6	−4	−6

Table 4.6
Deviation of the Scattered Radiation Intensity from the Normal over the USSR (in %)

Year	Month												Year
	I	II	III	IV	V	VI	VII	VIII	IX	X	XI	XII	
1964	10	7	12	22	17	6	19	6	3	9	11	25	14
1965	20	14	24	17	22	25	12	6	8	9	11	12	14

remained at most latitudes for about two years. It should be noted that the above data for the Soviet Union exhibit a longer period of considerable reduction in the radiation after the eruption of the Agung volcano. It is possible that only relatively fine particulates reached the middle latitudes of the Northern Hemisphere and it took them longer to settle. Dyer and Hicks present data from various sources on the properties of the aerosol cloud that originated after the Agung eruption. The average altitude of the cloud was evidently 15–20 km, the average radius of the aerosol particles varied from 0.5–1.0 μm in the first months to 0.1–0.15 μm a year later. Dyer and Hicks believe that the aerosol cloud formed at a height of 22–23 km and then its particles began settling at speeds appropriate to their sizes.

Taking into account a strong dependence of the radiation regime on volcanic eruptions, we may suppose that an increase in the radiation at the end of the nineteenth century resulted from the removal from the atmosphere of the aerosol particles produced by the Krakatoa eruption. A subsequent reduction in the radiation took place after the eruptions of Mont Pelée and other volcanoes in the early twentieth century [anomalies start decreasing on the curve of secular variations in the radiation (Figs. 4.17, 4.18) before these eruptions took place, which is explained by the 10-year smoothing effect]. The 1915–1920 increase in radiation is apparently explained by an increase in transparency that resulted from the settling of the aerosol after the Katmai eruption. After this, there were no large eruptions for a long time.

The problem of how changes in radiation after certain intensive eruptions affect the thermal regime have been dealt with in a number of empirical studies. As has been established, the mean temperature near the earth's surface decreased by a few tenths of a degree within a period of several months to several years after large eruptions (Humphreys, 1929; Mitchell, 1961, 1963; Angell and Korshover, 1977). Let us consider the effect of the radiation change due to volcanic eruptions on temperature. It is possible to calculate the effect according to a theoretical scheme, taking into account the heat conductivity and the thermal capacity of ocean water. Since we are interested only in an estimate of this effect, we shall use a simple empirical relationship between radiation and temperature fluctuations.

Let us assume that the rate of the mean temperature variation at the earth's surface is proportional to the difference between the temperature at a given moment (T) and the temperature corresponding to stationary conditions (T_r), i.e.,

$$dT/dt = -\lambda(T - T_r), \tag{4.3}$$

where λ is a coefficient of proportionality. Designating the temperature at an initial moment by T_1, we obtain from (4.3)

$$T - T_r = (T_1 - T_r)e^{-\lambda t} . \tag{4.4}$$

For rough estimation of λ, we use data on the annual variations in solar radiation and temperature for the Northern Hemisphere, neglecting the interaction of climatic conditions in the Northern and Southern Hemispheres.

Since the ratio of the solar radiation at the outer boundary of the atmosphere in the Northern Hemisphere during the warm seasons (April–September) to the average annual radiation is 1.29, we find from the relationship of temperature to radiation changes (Table 3.6) that this change in radiation could lead to an $\sim 40°$ temperature rise in the case where the heat inertia is absent. The observed difference between the mean temperatures in the Northern Hemisphere for the warm season and for the entire year is 3.5°. In this case, if $T - T_1 = 3.5°$ and $T_r - T_1 = 40°$ for $t = \frac{1}{4}$ of the year, we find $\lambda \cong 0.4/\text{yr}$ according to Eq. (4.4).

Taking this result into account, let us determine the mean temperature variations throughout the year after a volcanic eruption that has caused the direct radiation to decrease by 10% on the average over the year. In this case the decrease in the total radiation is 1.5% and T_r drops by $\sim 2°$. We find from (4.4) that the temperature change after the eruption appears to be a few tenths of a degree. This result agrees well with the mean annual temperature anomaly after an intensive explosive eruption.

A more detailed calculation of temperature variations after volcanic eruptions of an explosive nature has been carried out by Borzenkova (1974) who also used for this purpose a semiempirical model of the thermal regime of the atmosphere and Eq. (4.4) involving the mean temperature change near the earth's surface with time. Borzenkova has concluded that after a volcanic eruption the temperature drop in the Northern Hemisphere becomes more pronounced with increasing latitude in both the warm and cold seasons of the year.

Such calculations, of course, present a very schematic picture of the effect of volcanic eruptions on the thermal regime of the atmosphere.

Thorough empirical investigations of the air temperature change after intensive volcanic eruptions of an explosive nature carried out by Pokrovskaya (1971) and Spirina (1971) have revealed more complex regularities in the effect of a single eruption on the thermal regime of the atmosphere. They have established that after intensive eruptions the air temperature falls significantly for a few years during the warm seasons. In the Northern Hemisphere this temperature drop reaches it maximum in the northern part of the middle latitudes. During the cold seasons the temperature change acquires a more complex character. The temperature usually decreases in the polar zone but it often increases in the middle latitudes. As a result, the mean annual temperature drops considerably more in high latitudes than in mid-latitudinal belts.

According to Pokrovskaya, these regularities in the air temperature variations after eruptions are explained by the prevailing influence of the radiation factors on the climatic conditions of the warm seasons when an increase in the optical depth of the aerosol layer adds to the temperature drop in the polar zone. During the cold seasons the temperature distribution depends mostly on the fluctuations in the atmospheric circulation that are caused by variations in the heat income at various latitudes.

Tables 4.3 and 4.5 demonstrate that in the middle latitudes the relative effect of the aerosol particles on radiation is less intensive in the warm seasons than in the cold ones. The cause is evidently the increase in the optical depth of the aerosol layer with a decrease in the average altitude of the sun.

Since in the middle latitudes the amount of radiation is considerably less in the cold seasons than in the warm period of the year, the annual change in the absolute values of radiation variations appears to be much less than that of the relative values given in Tables 4.3 and 4.5. All the same, a comparatively large decrease in the radiation during the cold period cannot but strongly affect other components of the heat balance of the earth–atmosphere system. It is likely that this effect reveals itself mostly in the variations of the heat exchange in the upper oceanic layers whose large thermal inertia results in the summer temperature change described by Pokrovskaya and Spirina.

To determine the effect of volcanicity on long-term (of the order of a few decades) temperature variations we should look at the data presented by Lamb (1970a), who compared the temperature anomalies over different periods of time with an index describing the average reduction of the atmospheric transparency after eruptions. Lamb expressed this index in relative units that define the effect of volcanic eruptions on the transparency of the atmosphere in terms of the effect of the Krakatoa eruption that took place in 1884.

A correlation coefficient of -0.94 has been obtained between Lamb's index and the temperature anomalies in the Northern Hemisphere (from the equator to 60°N) averaged over decades from 1870 to 1959. This confirms a close relationship between volcanicity and temperature variations. According to Lamb, this relationship failed at the end of this period. The reasons for this failure will be considered in Chapter 5.

The relation between the amount of sea ice along the shoreline of Iceland and the index of atmospheric opacity suggested by Lamb is of certain interest in view of the above conception of the nature of modern climatic changes. The correlation coefficient between the above parameters appears to be 0.61 for 10-year intervals from 1780 to 1959. This seems to be rather high, since estimates of the characteristics of the processes studied were fairly rough, especially for the first part of the given period.

In elucidating the mechanism of the influence of volcanic eruptions on the climate we should dwell on the dependence of climatic conditions on the atmospheric aerosols. In addition to water droplets and ice particles in clouds and fogs, the atmosphere contains a large number of suspended solid and liquid particles of different chemical composition. These particles range in size from 10^{-2} to $10^{-6}-10^{-7}$ cm. Most of the atmospheric aerosol comprises the so-called "large" particles with a radius from 10^{-5} to 10^{-4} cm and "giant" particles over 10^{-4} cm in radius.

The available estimates show that from 800 to 2200 million tons of the particulate matter forming the aerosol particles enter the atmosphere annually from natural sources. About 200–400 million tons are formed by man-made pollution. The aerosol particles include airborne soil particles, sea-salt particles from the ocean surface, soot and ashes from forest fires and combustion gases, as well as particles formed by chemical reactions of sulfur dioxide, hydrogen sulfide, ammonia, and other gases entering the atmosphere, from the earth's surface (SMIC, 1971).

The great majority of atmospheric aerosol particles is concentrated in the lower layers of the troposphere and the residence time of separate aerosol particles there is relatively short. The larger particles fall out under the influence of gravitational forces more rapidly than the smaller ones, which are transported downward to the earth's surface by the air currents and atmospheric precipitation that play an important part in removing aerosols from the atmosphere. According to the available estimates, the average residence time of aerosol particles in the troposphere is about 10 days. Taking this into account, we find that the atmosphere contains approximately 30–70 million tons of aerosol.

The aerosol burden in the stratosphere is considerably less. However, the aerosol particles there have a much longer lifetime as compared with the troposhere. The majority of the stratospheric particles range from 1 to 0.1 μm in radius (large particles). They are less affected by gravitational forces and fall out relatively slowly. Since there are weak vertical air motions and no precipitation, the aerosols have a residence time in the stratosphere of a few months to a few years.

The direct observations of the stratopheric aerosols that were started by Junge (1963) have shown that their typical forms are droplets of H_2SO_4 and, to a lesser extent, salts of H_2SO_4, mainly represented by ammonium compounds. The majority of stratospheric aerosol particles are concentrated in a layer several kilometers thick whose center is at an altitude of 18–20 km. This layer is called the sulfate layer or the Junge layer.

According to Junge, the aerosol particles of the sulfate layer generally originate from SO_2 entering the stratosphere from the lower layers of the atmosphere. Sulfur dioxide interacts with atomic oxygen by photochemical reaction and forms SO_3, which turns into droplets of H_2SO_4 by interaction with water vapor.

Studies of circulation processes in the stratosphere have shown that, if a source of stratospheric aerosol is in the extratropical latitudes, the aerosol spreads rapidly within the hemisphere but disperses only slowly into the other hemisphere. If the aerosol source is near the equator, it spreads through both hemispheres (Karol, 1972).

The particles of stratospheric aerosol gradually settle, both under the influence of the gravitational force and as a result of large-scale air motions that transfer the particles to the troposphere where they are rapidly washed out by precipitation. The second of these mechanisms for removing aerosols from the stratosphere is evidently important for large particles and the first for giant particles, whose number in the stratosphere is insignificant due to their rapid removal.

According to Karol's data (1973), large particles remain in the stratospheric layer at a height of 20 to 30 km for an average of 20 to 40 months. In the tropopause this term contracts to 6–20 months. Both periods can vary considerably according to the intensity of air exchange between the stratosphere and the troposphere.

The study of the dependence of vertical radiation fluxes in the atmosphere on aerosol concentration is of great importance in clarifying the effect of atmospheric aerosols on climate. Starting with the above-mentioned works of Humphreys, this dependence has been treated by many authors. Humphreys's conclusion concerning a usually slight influence exerted by atmospheric aerosols on longwave radiation has been confirmed by subsequent studies.

Moreover, it has been established that atmospheric aerosols can appreciably change the flux of shortwave radiation due to backscattering and absorption of the radiation by the aerosol particles. As mentioned above, in the process of backscattering of radiation by particles whose sizes range from a few tenths of a micron to several microns, the scattering is strongly increased in the direction of the incident ray and, as a result, the aerosols affect the direct radiation more strongly than the total radiation. The question relating to the absorption of shortwave radiation by the aerosol particles is less thoroughly treated than that of backscattering since there is a shortage of experimental data on absorption coefficients. The available data show (Gaevskaya, 1972) that the reduction of the direct radiation flux due to absorption by the aerosol particles is obviously less than that resulting from backscattering, although both effects can be quantitatively comparable.

We can present some estimates of the effect of the atmospheric aerosol on the reduction of the shortwave radiation flux. The simplest method of obtaining such an estimate is based on the comparison of the average climatological value of aerosol reduction of the shortwave radiation flux with the average mass of atmospheric aerosols.

According to Pivovarova's (1968) and other authors' data, we shall assume that the average aerosol reduction of the direct radiation comes to $\sim 10\%$. It follows from the above data that the mean aerosol mass in the atmosphere is ~ 50 million tons or 10^{-5} g cm^{-2}. Roughly half of this amount refers to the giant particles whose number is insignificant; therefore they affect the radiation process slightly. In this case an aerosol mass of 0.5×10^{-6} g cm^{-2} reduces the direct radiation by 1%.

If this reduction was explained only by backscattering of radiation by the aerosol particles, the corresponding decrease in the total radiation for average conditions would come to $\sim 0.15\%$. If the direct radiation reduction resulted entirely from its absorption by the aerosol particles, the decrease of the total radiation evidently would have been 1%. Let us assume that both effects have a comparable influence on the reduction of the direct radiation. In this case the aerosol mass decreasing the total radiation by 1% should be $\sim 10^{-6}$ g cm^{-2}. This estimate is, of course, very rough.

In a number of studies calculations have been made concerning the effect of the aerosol mass on the reduction of the total radiation flux resulting from backscattering of radiation by the aerosol particles. Making the first calculation of this type, Humphreys (1929) found that the radiation decreases by 1% with an aerosol mass $M = 0.6 \times 10^{-6}$ g cm^{-2}. Similar calculations have been made by Novosel'tzev on the basis of a scheme of repeated scattering suggested by Sobolev and by Vinnikov on the basis of a scheme of single scattering of Schifrin. They obtained M values of 0.4×10^{-6} and 0.6×10^{-6} g cm^{-2}, respectively. A somewhat greater value has been obtained by Barrett (1971) who found $M = 10^{-6}$ g cm^{-2} with scattering by the aerosol particles at low concentrations.

The value of M can also be obtained by calculation of backscattering by the aerosol particles based on the model of Yamamoto and Tanaka (SMIC, 1971). According to these data M appears to be 1.3×10^{-6} g cm^{-2}.

The effect of the aerosol mass on the reduction of shortwave radiation can be estimated on the basis of the Ångström (1962) formula, which relates the albedo of the earth–atmosphere system to the index of optical opacity, taking into account an average aerosol mass. According to this method $M = 0.6 \times 10^{-6}$ g cm^{-2}.

We shall not dwell on the results of similar calculations but mention that the average value of M evidently lies in the range of 0.4 to 1.3×10^{-6} g cm^{-2}. This conclusion coincides almost exactly with the results obtained in the study by CIAP (Grobecker *et al.*, 1974).

Considering that the average M is far less than the variability of the aerosol mass, we can conclude that fluctuations in the atmospheric aerosol mass with time and space can noticeably change the flux of shortwave radiation incident on the earth's surface. These variations seem to exert a significant influence on the thermal regime of the atmosphere.

The question of the relationship of atmospheric aerosol variations and volcanic activity is worth noting. An indirect evidence of the presence of such a relationship lies in the above data on sharp fluctuations in the direct solar radiation after intensive volcanic eruptions. These data reveal considerable changes in the aerosol concentration after the explosive volcanic eruptions.

There are also direct data on considerable growth of stratospheric aerosol mass after intensive eruptions. Measurements by high-altitude planes have shown that the aerosol concentration in the stratosphere of the Northern Hemisphere varied greatly during the 1960s. This change seems to be explained by the effect of the Agung volcano eruption in 1963 and several subsequent eruptions of other volcanos (SMIC, 1971).

The notion of the physical nature of contemporary climatic change presented here was set forth in the author's works published in 1967–1974.

Many other investigations were published in subsequent years with similar conclusions about the causes of modern climatic change. The studies of Pivovarova (1977a,b) are among the works devoted to this problem. They deal with secular variations in atmospheric transparency and show a significant role for volcanic eruptions in this process. Research carried out by Karol (1977) and Karol and Pivovarova (1978) is concerned with fluctuations in the direct solar radiation and the air temperature at the earth's surface and with their dependence on the stratospheric aerosol.

In the first of these studies aircraft data on stratospheric aerosol at a height of 14–20 km in the Northern and Southern Hemispheres over the 1960–1973 period was compared with the anomalies of direct solar radiation obtained from a number of actinometric stations covering the middle latitudes of the Northern Hemisphere.

A close relationship was revealed between the two parameters, corroborating the conception that fluctuations in stratospheric aerosol are extremely important in terms of climate and its change. This relationship was used to evaluate the effect of stratospheric aerosol fluctuations due to volcanicity on the thermal regime of the lower atmosphere. The results appear to be similar to the temperature change observed after volcanic eruptions.

The second study is concerned with a comparison of US balloon reports on the atmospheric aerosol with fluctuations in the transparency coefficient based on the direct solar radiation measured at a number of Soviet actinometric stations.

As a result of this investigation it was found that the actinometric observations of the direct solar radiation, especially on mountains and at far-off stations, reveal the influence of major volcanic eruptions on the atmospheric transparency.

American scientists have also contributed to the study of the dependence of climate on volcanic eruptions.

Oliver (1976) has calculated air temperature variations in the Northern Hemisphere up to 1968 using a model that allowed him to determine the temperature variations caused by a change in atmospheric transparency after volcanic eruptions. His calculations show that such an approach can be used to describe about 70% of the variance of temperature series. The remainder can be considered as a characteristic of meteorological noise. We believe that if, in addition, anthropogenic factors of climatic change are taken into account (see Chapter 5), the noise will be less pronounced.

The same problem has been treated by Sagan and co-workers (Pollack et al., 1975, 1976; Baldwin et al., 1976), who consider in detail the effect of the debris of volcanic eruptions on the radiation regime of the atmosphere.

According to observational data, the ejected material seems to comprise solid particles which fall out fairly quickly (the residence time of finer fragments extends to several months), and droplets of sulfuric acid produced from sulfur oxides (with a longer residence time).

In studies of the effects of the Agung eruption on the climate, observational data were noted that showed that the lower stratospheric layers were appreciably heated in the first months after the eruption. According to calculations, this could have been a result of the absorption of longwave radiation by igneous particles. A certain temperature rise may also affect the troposphere immediately after the eruption when the largest particles of igneous aerosol remain in the air.

Later on this effect disappears and the troposphere and stratosphere become cooler. Taking into account the effect of the inertia of the earth–atmosphere system, it appears that the mean temperature in the lower air layer drops by several tenths of a degree after a major volcanic eruption. This conclusion agrees reasonably well with the observational records.

Considering the effect of volcanic eruptions and CO_2 growth on the thermal regime of the atmosphere, variations in the mean temperature of the Northern Hemisphere throughout the last century have been calculated. The results proved to be close to these based on observational data.

The studies in question also treat the problem of how much rising volcanicity contributed to the progression of cooling a few centuries ago, during the period often referred to as the Little Ice Age. The calculations show that the contribution might be very significant.

Hansen et al. (1978) have analyzed similarly the effects of the Agung eruption that occurred in 1963. They concluded that all the quantitative estimates obtained of the thermal regime change after the eruption agree perfectly with the observational data.

Worth noting is a study by Miles and Gildersleeves (1977) containing an empirical analysis of the relationships between the mean air temperature variations in the Northern Hemisphere and climate-forming factors. The authors concluded that the mean hemispheric temperature variability over the last hundred years is mainly related to fluctuations in the atmospheric opacity due to volcanic activity and to a rise in the CO_2 concentration in the atmosphere. With these two factors in mind, it is possible to describe 65% of the temperature variability. The amount will increase to over 80% if a third factor, i.e., the variation in polar ice area, is taken into consideration.

It has been established by Robock (1978) on the basis of a similar analysis that volcanic activity essentially affects the mean air temperature in the Northern Hemisphere whereas solar activity, which is characterized by the number of solar spots, does not exert such an influence.

Summarizing these ideas, we can set forth a conclusion concerning the reason for the present climatic changes.

The warming that peaked in the 1930s evidently resulted from a stratospheric transparency increase that enhanced the solar radiation flux coming into the troposphere (the meteorological solar constant). Consequently, the global mean air temperature near the earth's surface began to rise.

Modifications in the air temperature in different seasons and latitudes were dependent on the optical depth of the stratospheric aerosols and on the shift of the polar sea ice boundary. A retreat of Arctic sea ice due to the warming gave rise in addition to an appreciable increase in the air temperature in the cold seasons at high latitudes of the Northern Hemisphere.

These conclusions are corroborated by calculations based on the model of the thermal regime of the atmosphere, the results being in good agreement with the observational data.

Fluctuations in the stratospheric transparency during the first half of the twentieth century have probably been due to variations in volcanic activity and, in particular, to variations in the release of ejected materials in volcanic eruptions, especially of sulfur dioxide into the stratosphere.

A similar physical mechanism was at work in the second half of the twentieth century, when the climate was particularly affected by the Agung eruption in the mid-1960s.

In addition to the effects of volcanicity, climatic conditions of the twentieth century were under the influence of man's economic activity, which, in particular, brought about a rise in the CO_2 concentration in the atmosphere. The latter will be considered thoroughly in Chapter 5.

Undoubtedly these reasons for contemporary climatic changes cover only their major features.

In line with the above general regularities of climatic change there are also many peculiarities in short-term climatic variations and in variations in separate geographical regions.

We consider such variations of climate to be mainly determined by fluctuations in the atmospheric and hydrospheric circulation, which to a great extent are of a random nature.

Chapter

5

Man's Impact on Climate

The heat balance of the earth's surface

Man's economic activity affecting climatic conditions over limited regions is usually associated with changes in the heat (energy) balance of the earth's surface. It is necessary to study these changes to understand the physical mechanism of man's impact on local climate.

The equation for the heat balance of the earth's surface used in Chapter 3 has the form

$$R = LE + P + B, \qquad (5.1)$$

where R is the radiation balance of the earth's surface, LE the heat expenditure for evaporation (L is the latent heat of vaporization and E the rate of evaporation or condensation), P the turbulent flux of heat between the earth's surface and the atmosphere, and B the heat flux between the earth's surface and lower layers of water or soil.

This equation can be rewritten as

$$R_0 - 4\delta\sigma T^3(T_w - T) = \rho C_p D(T_w - T) + LE + B. \tag{5.2}$$

$R_0 = Q(1 - \alpha) - I_0$, where Q is the total shortwave radiation incoming to the earth's surface, α the albedo of the earth's surface, and I_0 the net long-wave radiation determined taking into account the air temperature; $4\delta\sigma T^3(T_w - T)$ is the difference in net longwave radiation fluxes determined from the earth's surface temperature T_w and the air temperature T (where δ is a factor indicating the difference in properties of the emitting surface and of blackbody and σ is the Stefan constant); and $\rho C_p D(T_w - T)$ is the sensible-heat flux from the earth's surface to the atmosphere (where ρ is the air density, C_p the heat capacity of the air at constant pressure, and D the integral coefficient of turbulent diffusion).

All the components of this equation depend strongly on the state of the earth's surface, in particular, on its reflectivity (albedo), the degree of moistening, the presence and properties of vegetation, roughness, heat conductivity, and heat capacity of the upper soil layers. Typical albedo for various kinds of land surface are given in Table 5.1 (Budyko, 1971).

Calculations and observations show that, other things being equal, variations in the earth's surface albedo within the limits shown in Table

Table 5.1
Albedo of Natural Surfaces

Type of surface	Albedo
Snow and ice	
Fresh, dry snow	0.80–0.95
Clean, moist snow	0.60–0.70
Dirty snow	0.40–0.50
Sea ice	0.30–0.40
Bare soil	
Dark soil	0.05–0.15
Moist, grey soils	0.10–0.20
Dry loam or grey soils	0.20–0.35
Dry, light, sandy soils	0.35–0.45
Fields, meadows, tundra	
Rye and wheat fields	0.10–0.25
Potato fields	0.15–0.25
Cotton fields	0.20–0.25
Meadows	0.15–0.25
Dry steppe	0.20–0.30
Tundra	0.15–0.20
Arboreal vegetation	
Coniferous forest	0.10–0.15
Deciduous forest	0.15–0.20

5.1 may have a considerable effect on the value of the radiation balance of the earth's surface and, therefore, on climatic conditions in a given region. Comparatively small changes in the albedo (of several percent) often lead to an increase or decrease in the temperature of the lower air layer that can be detected from meteorological observations.

The climate of the lower air layers is similarly affected by variations in the heat expenditure for evaporation, which is dependent on the amount of soil moisture.

In studying this relationship, we can use the following approximations (Budyko, 1971):

$$E = E_0 \qquad \text{at} \quad w \geq w_0$$

and

$$E = E_0 \, w/w_0 \qquad \text{at} \quad w < w_0, \tag{5.3}$$

where E_0 is the potential evaporation from soil that is sufficiently moist (potential evapotranspiration), w the moisture of the upper soil layer, and w_0 the critical value of the soil moisture.

The w_0 value is usually equal, for the upper soil layer of 1 m depth, to the amount of available moisture corresponding to a 10–20 cm layer, depending on vegetation cover and its properties. Since the moisture of the upper soil layer varies over a wide range, it greatly affects the LE component and, therefore, the climate of the lower air layer.

The local climatic conditions also depend on the aerodynamic roughness of the earth's surface. For an even snow or ice surface the roughness is usually of the order of 10^{-3}–10^{-2} cm. For an even meadow covered by low grass it is 0.1–1 cm. For high grass or agricultural plants like potatoes the roughness increases up to 1–10 cm. For bushes and forest it may be 10–100 cm. The roughness influences wind speed in the lower air layer and also the coefficient of turbulent diffusion, which increases with increased roughness. Variations in the turbulent exchange intensity lead to variations in the turbulent heat flux between the earth's surface and the atmosphere and they affect the heat expenditure for evaporation. This influences the temperature of the earth's surface and the lower air layers.

Although the thermal properties of soil can vary within wide limits, related variations in the B values usually do not affect local climate since in most cases the heat flux directed into the soil is less in absolute value than other heat balance components. Nevertheless, this heat flux may be of great importance when comparing the climatic conditions of an island and a reservoir surrounding it or of a lake and shore regions, because under these conditions the differences among the B values can be significant.

It should be emphasized that the effects of variations in heat balance components on the climate of the lower air layers depend strongly on the area of the territory in which these variations are observed. For small territories, on the scale of several square meters, climatic changes are insignificant but they increase rapidly as the territory grows. For territories of the order of several square kilometers and more, the dependence of climatic variations on the scale is weaker. A similar regularity also exists with respect to the thickness of the air layer in which meteorological elements vary due to changes in the state of the earth's surface.

Of the various kinds of impact man makes on local climate, the effects associated with changes in the state of vegetation and water balance are the most widespread. Specific climatic conditions are also observed in territories occupied by more or less large cities and towns.

Effects on vegetation cover

Man's impact on climate became apparent hundreds of thousands of years ago through deforestation.

Since antiquity, fire has been an important factor in human influence on the environment, causing deforestation over large territories. Forest and steppe fires have long been widely used as a means of hunting large animals. Until recently this method was used by Australian aborigines who destroyed vegetation in the areas of tens of square kilometers when hunting (Dorst, 1965). Similar methods of hunting were used by people in the Upper Paleolithic and even earlier. This may explain the rapid deforestation in many regions after the appearance of modern man.

In the Neolithic epoch, when cattle breeding and agriculture became the major areas of man's economic activity, the burning out of vegetation took place on a larger scale. Thus the area of pasture and arable land was expanded at the expense of forests. Also, the use of slash—burn agriculture, based on the felling of trees and their burning, gave rich harvests even with superficial cultivation of the soil, because of its ash fertilization.

With the use of this method of agriculture the fertility of soil decreases rapidly. Therefore, after several years (sometimes in one or two years), new forest areas must be felled. This method can be used if there are vast thinly populated forest areas. In the past it was used in many countries at middle latitudes and it is now used in some developing countries at tropical latitudes.

The wide use of burning out of vegetation over a large area of land resulted in drastic changes in natural conditions, including flora, fauna, soil, climate, and the hydrological regime. The total effects of the destruction of forests are difficult to estimate because deforestation began very long ago. In many cases vegetation destroyed by man is never renewed, even if systematic burning out is stopped.

In some regions forests were destroyed for their wood. In many territories natural vegetation cover was greatly affected by cattle pasturing, frequently carried out without consideration of the renewal of the vegetation. In forest regions with arid climates goats and other domestic animals ate the foliage of young trees; this led ultimately to deforestation. Excessive pasturing of cattle destroyed vegetation in dry steppes and savannas, which then frequently turned into semideserts or deserts.

Deforestation resulted in increased wind speed at the earth's surface, in changes of temperature and moisture diurnal variations in the lower air layer, and in changes in soil water content, evaporation, and river runoff. In comparatively arid zones, deforestation is frequently accompanied by intensified dust storms and soil erosion. These considerably affect natural conditions in these territories.

In addition, deforestation over vast areas exerts some influence on large-scale meteorological processes.

Decreased roughness of the earth's surface and changed evaporation in deforested territories can affect precipitation patterns. This effect is insignificant if forests are replaced by other types of plant cover (Drozdov and Grigor'eva, 1963). Precipitation may be affected more appreciably by the complete destruction of vegetation in some territory, as has repeatedly occurred in the past due to man's economic activity. For example, this took place in some mountain regions with weakly developed soil layers. Under these conditions, the soil layer is rapidly eroded if forest does not protect it. As a result, the further existence of mature plant cover becomes impossible. A similar situation arises in some regions of dry steppes, where natural vegetation cover destroyed by excessive grazing of cattle is not renewed and these regions turn into deserts.

The earth's surface, without vegetation, is intensely heated by solar radiation. This leads to a decrease in relative air humidity and to a rise in the level of condensation that may decrease precipitation. This mechanism seems to explain some cases where natural vegetation in dry regions is unrenewed after being destroyed by man.

The climate can be somewhat affected by variations in the earth's surface albedo as conditioned by deforestation. Table 5.1 shows that dry steppe replacing forest increases the albedo by about 0.1. If the steppe, in its turn, becomes a desert, the albedo may increase again by approximately the same amount. This may result in an air temperature drop at the earth's surface due to decreased radiation balance if other heat balance components do not change. The latter condition, however, is usually not fulfilled. When vegetation is destroyed the heat expenditure for evaporation decreases, resulting, as noted above, in elevation of the earth's surface temperature and consequently of the air temperature.

The destruction of vegetation over vast territories may decrease the global radiation balance of the earth—atmosphere system and, at least, change the global climate toward cooling. This will be discussed below.

The planting of forests is also accompanied by changes in meteorological conditions. Of the various kinds of forest plantings, field-protecting forest belts exert the greatest influence on the climate of the air layer near the earth's surface. These plantings are widely used as means of land improvement. Forest belts are usually from several meters to 50 meters wide. They border square or rectangular fields with sides several hundred meters to 1 to 2 kilometers long. Forest belts are often grown in arid regions where they help to maintain more favorable moisture conditions in farm lands.

The wind-breaking effect of forest belts, which decreases the mean wind speed over the fields between the belts and the rate of turbulent exchange near the earth's surface, is the principal way that they influence the meteorological regime of the lowest air layer. The attenuation of turbulent exchange in the fields between the belts can be attributed to the breaking up of vortices moving near the earth's surface as they pass through a forest belt. This substantially decreases the intensity of eddy motions in the flow (Yudin, 1950).

The turbulence directly affects the development of two meteorological phenomena, the blowing of snow from fields and the formation of dust storms. A decrease in turbulent exchange near the earth's surface prevents or weakens dust storms and aids in maintaining snow on the fields between the belts. It is also important for preserving moisture reserves in soil during the warm half-year.

The potential evaporation, among other meteorological factors, depends on the intensity of turbulent exchange in the lowest air layer. Calculated results (Budyko, 1971) have shown that the potential evaporation decreases by approximately 10% with a 20% decrease in the mean value of the turbulent exchange coefficient for the lowest air layer.

Besides decreasing potential evaporation, protective forest belts aid in increasing snow reserves on fields and the amount of precipitation. The influence of all these factors leads to a significant increase in soil moisture in fields protected by forest belts.

As noted above, numerous observations have shown that the runoff of snow water decreases considerably in fields with protective forest belts. This can be attributed to a change in the distribution of snow cover on protected fields as compared with unprotected fields. In protected fields, a decrease in wind speed and turbulent exchange in the lower air layers create conditions for a comparatively uniform distribution of snow cover, whereas in open fields, a significant portion of snow is blown away into ravines and other topographic depressions, and drains away after melting. Moreover, under forest belts the soil permeability increases, providing somewhat greater retention of snow water in protected fields as compared with open ones; this also decreases the spring runoff of snow water.

The water balance of soil protected by forest belts can be somewhat affected by a change in evaporation and in the amount of precipitation due to variable vertical velocities in the atmosphere over the forest belts.

In order to estimate the total effect on soil water balance of variations in the above-mentioned meteorological conditions, we may use heat- and water-balance equations, taking into account the values of the integral diffusion coefficient, runoff, and precipitation under the conditions of planted protective forest belts (Budyko, 1956). The results obtained show that in the fields protected by forest belts, soil moisture increases considerably and evaporation grows somewhat. This increase in soil moisture differs during various seasons, depending on changing turbulent exchange, runoff, and precipitation.

An appreciable increase in available soil moisture and some increase in total evaporation can substantially improve the crop yield in temperate climatic conditions.

Protective forest belts, in addition to influencing soil-water balance, also play an important role in abating dust storms, which drastically damage the soil cover in arid regions. This was clearly manifested in the winter of 1968–1969 when powerful dust storms occurred in the southern part of the USSR European territory. Inspection showed that winter crops on protected fields suffered significantly less damage than those on unprotected fields.

Protective forest belts are also used in regions of sufficient moisture, where they increase the mean surface temperature during the warm season by attenuating turbulent mixing. Under such conditions they exert a favorable influence on the development of thermophilic crops and accelerate the maturation of many agricultural plants (Gol'tsberg, 1952).

Without dwelling on other methods of changing the meteorological regime, we should note that man's impact on vegetation cover changes local climatic conditions in the lower air layer within certain limits.

Impact on the water regime

The use of artificial irrigation is one of the ways man affects climate. In arid regions irrigation has been utilized for thousands of years, beginning with the most ancient civilizations in the Nile valley and the Tigris–Euphrates area.

Irrigation drastically changes the microclimate over the irrigated fields. A significant increase in heat expenditure for evaporation leads to a decreased surface temperature that causes a reduction in temperature and a rise in the relative humidity of the lower air layer. This variation in the meteorological regime, however, disappears quickly beyond the irrigated fields. Therefore, irrigation results only in variations of the local climate and affects large-scale meteorological processes but slightly.

Let us consider in detail the physical mechanism of variations in the meteorological regime due to irrigation (Budyko, 1956).

As a result of irrigation of arid steppes, semideserts, and deserts, the radiation balance increases significantly. This increase may be comparable with the value of the initial balance. On one hand, an increased radiation balance can be attributed to increased absorption of shortwave radiation due to reduction in the albedo, which is much less for moist soil covered by more or less abundant vegetation than for deserts or semideserts. On the other hand, lowering the temperature of the underlying surface and raising the humidity of the lower air layer by irrigation decrease the effective longwave radiation. This also raises the radiation balance.

Irrigation under arid climatic conditions causes a sharply increased heat expenditure for evaporation, the magnitude of which depends mainly on irrigation rates.

As a rule, the increase in heat expenditure for evaporation under ordinary irrigation rates exceeds the increase in the radiation balance. Therefore, if irrigation rates are sufficiently large, the turbulent heat transfer is noticeably decreased and reaches negative values that correspond to the direction of mean turbulent heat flux from the atmosphere to the underlying surface. This causes the appearance of daily temperature inversions.

Thus, irrigation under arid climatic conditions significantly decreases both the turbulent heat flux (which can even change its direction) and the heat flux transmitted by longwave radiation. For irrigation of sufficiently large areas this may lead to appreciable variations in air-mass transformation over a given territory.

Variations in the heat balance components due to irrigation can be estimated from the data of observations at Pakhta-Aral (Central Asia). The results of these observations as depicted in Fig. 5.1 allow us to compare the heat balance components of an irrigated oasis and the semidesert surrounding it for summer conditions.

As seen in Fig. 5.1, the radiation balance R increases noticeably in the oasis as compared with the semidesert, and the heat expenditure for evaporation from the irrigated fields is great (in the semidesert, evaporation over the period in question equaled zero). Therefore, the turbulent heat flux P in the semidesert is much greater during the daytime than in the oasis and is opposite in sign. The heat flux B in the soil under the given conditions changes comparatively little as a result of irrigation.

As noted above, the thermal regime is considerably affected by irrigation. In deserts and arid steppes, because of low heat expenditure for evaporation, solar radiation absorbed by the earth's surface is basically consumed in heating the atmosphere through turbulent heat transfer and longwave radiation. Under these conditions, very high surface temperatures are observed. To assess these temperatures, we may use the heat balance equation (5.2), from which it follows that

$$T_w - T = \frac{R_0 - LE - B}{\rho C_p D + 4\delta\sigma T^3}. \tag{5.4}$$

Since LE and B are much less than R_0 under these conditions, $T_w - T$ can reach high values in daytime. From (5.4) it follows that $T_w - T$ may be as high as $10\text{--}20°$.

Large values of $T_w - T$ correspond to intensive heat flux from the earth's surface to the atmosphere. As a result of heating of the lower air

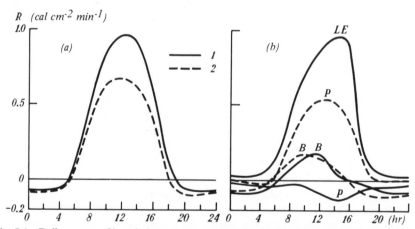

Fig. 5.1 Daily course of heat balance components (1) in an irrigated oasis and (2) in a semidesert. (a) Daily course of radiation balance; (b) daily course of heat expenditure by evaporation (LE) turbulent heat flow (P), and heat exchange in soil (B).

layer, its temperature rises and the relative humidity drops. Reduced relative humidity, in its turn, aids in decreasing precipitation.

The soil-water balance essentially changes as a result of irrigation of arid zones. Irrigation causes a sharp increase in evaporation, which is equal to the irrigation rate minus the loss of irrigation water by infiltration. Consequently, the heat loss for evaporation also increases, leading to a considerable lowering of the earth's surface temperature. Taking into account that the air temperature in irrigated zones varies considerably less than the surface temperature, one may derive a formula for determining variations in surface temperature for irrigated areas:

$$T_w - T'_w = \frac{(R_0 - R'_0) - (LE - LE')}{\rho C_p D + 4\delta\sigma T^3}, \tag{5.5}$$

where the values referring to irrigation are indicated by a prime. In deriving Eq. (5.5) it was assumed that $|B - B'| \ll |LE' - LE|$.

As indicated above, although the radiation balance of the earth's surface increases strongly in the course of irrigation due to a decrease in the albedo, the rise in heat expenditure for evaporation significantly exceeds the increase in the radiation balance.

As a consequence, the difference $T_w - T'_w$ is quite large for sufficiently high irrigation rates. The temperature of the earth's surface over an irrigated area under daytime conditions approaches the air temperature and in the case of abundant irrigation it is less than the air temperature.

We noted above that for irrigation of vast areas this leads to considerable changes in air-mass transformation. As a result, the temperature and humidity of the lower air layer changes over irrigated territories. Dry warm air entering from outside is humidified and cooled as it moves over an irrigated area.

Temperature and humidity variations of the air over irrigated areas can be attributed to the transformation of air masses arriving from outside at certain wind directions. The magnitude of these variations depends on the distance from the border of the irrigated oasis. The longer the distance, the greater the variations. In addition, temperature and humidity variations of the air due to irrigation depend on the following factors:

(1)   Irrigation rates and time intervals between irrigations. The greater the amount of water an irrigated field receives and evaporates over a given time interval, the greater the temperature and humidity differences between irrigated and nonirrigated fields.

(2)   Wind speeds and coefficients of turbulent exchange. These two closely related factors govern the thickness of the air layer in which temperature and humidity fluctuations occur. The temperature and humidity

distributions with altitude within the lower air layer chiefly depend on the same factors.

(3) Radiation properties of the underlying surface, mainly its reflectivity (albedo): This factor is discussed above.

To study empirically the effect of each factor separately is rather difficult. Therefore, it is more advisable to evaluate the temperature and humidity variations theoretically and to verify the results by data from stationary and field experimental studies (Budyko *et al.*, 1952).

On the basis of developed theory we can calculate the temperature and humidity fluctuations at different altitudes depending on water losses for evaporation, variations in albedo, rate of turbulent exchange, and distance from the oasis border. These calculations can be used for planning irrigation systems over previously nonirrigated areas.

Let us move on to the consideration of some empirical data.

Table 5.2 presents data of observations at meteorological stations located in a desert and in irrigated oases. The data correspond to a latitude of 42° and altitude of 100 m above sea level. Oases with a width not exceeding 3 km were considered small, the others large. As seen in Table 5.2, variations in temperature and humidity are the greatest in absolute value during the summer months when irrigation water discharge is the greatest. At this very time, the temperature drop at the standard level for ground meteorological observations (i.e., at 2 m above the earth's surface) caused by an increase in evaporation and transpiration amounts to 2.5–3°, while the absolute humidity at the center of a large oasis is roughly 5 mbar greater than in the desert. At lower altitudes, the absolute values of the differences increase.

In steppe zones, irrigation effects on air temperature and humidity are somewhat lesser. This is mainly due to lower additional water discharge for evaporation.

Table 5.2
Effects of Irrigation on Meteorological Regime

	IV	V	VI	VII	VIII	IX	X	XI
Temperature difference (°C):								
Small oasis–desert	0.0	−0.5	−1.6	−2.4	−2.5	−1.7	−1.4	−0.4
Large oasis–desert	−0.6	−1.1	−2.2	−3.1	−2.8	−2.3	−1.7	−0.8
Absolute humidity difference (mbar):								
Small oasis–desert	1.1	1.8	3.4	3.6	3.7	2.5	1.2	0.4
Large oasis–desert	0.4	1.8	4.2	5.4	5.4	3.6	1.6	0.8

It should be mentioned that irrigation of arid zones that reduces the temperature over irrigated fields increases the mean atmospheric temperature.

The temperature decreases in irrigated regions because of the growth of heat expenditure for evaporation but this growth for the entire earth is compensated for by an equal increase in heat income due to condensation. This heat is released in the atmosphere of other regions where the water vapor formed in the course of irrigation is condensed. In addition, the albedo of the earth's surface decreases significantly when arid steppes and deserts are irrigated. This increases the amount of radiation the earth–atmosphere system absorbs.

The large water reservoirs created during the construction of hydro-electric power stations are comparatively shallow. Therefore, their effect on climate is rather limited. This effect first decreases surface roughness, thus strengthening the wind speed, which is usually greater over water surface than over open flat territories. The wind speed increases mostly in autumn, when the water is warmer than the air and an intensive turbulent exchange develops over reservoirs. In spring, this effect is comparatively weak, both when reservoirs are covered with ice and immediately after it has melted, when the water is still cold. At this time, the wind speed over water basins is about the same as that over open flat territories. After the construction of a water basin, daily air temperature variations decrease, the radiation balance increases (as a result of decreasing albedo over a given area), and the mean annual evaporation, which during the year has a different distribution compared with land, grows.

Under conditions of excessive and sufficient moistening, there is not a great change in local climate following construction of water reservoirs. For example, any systematic temperature fluctuations are difficult to observe in the vicinity of the Rybinsk water storage basin at the stations located along its present shores.

Significantly larger climatic changes occur on the shores due to construction of reservoirs in regions with insufficient moistening (e.g., Tsimlyansk, Volgograd, and Bukhtarminsk reservoirs). The temperature along the shorelines of these basins is noticeably lower (by 2–3°) in the warm season than that in regions far from water reservoirs because of greater evaporation from the surface of the reservoir than from the surrounding land, where the evaporation rate is limited by low soil moisture and from which dry air comes to the shore of the basin. A daytime air temperature drop promotes the development of rather strong (up to 3–4 m sec^{-1}) breezes, extending vertically to several hundred meters.

Artificial reservoirs, like irrigation, cause a decrease in the earth–atmosphere system albedo and consequently an increase in the amount of absorbed radiation. Water basins thereby increase the mean temperature of the atmosphere.

Urban climate

Climatic conditions in cities differ somewhat from those in the surrounding regions and, the larger the area of the city, other conditions being equal, the greater the difference.

Probably climatic changes in large cities appeared hundreds of years ago. Thus Landsberg (1956) presented eyewitness evidence concerning severe air pollution in London in the seventeenth century that significantly attenuated solar radiation in the city as compared with rural areas.

The chief factors affecting the meteorological regime of a city are

(1) variations in the albedo of the earth's surface, which are usually less in built-up areas than in rural areas;

(2) a change in mean evaporation from the earth's surface, which is noticeably lowered within the boundaries of a city (although evaporation from roofs and pavements immediately after rainfall may exceed evaporation in a rural area);

(3) heat released due to various kinds of economic activity, the amount of which may be comparable to that of the solar radiation incident on the area of a city;

(4) increased roughness of the earth's surface within the boundaries of a city as compared with a rural area;

(5) atmospheric pollution by different solid, liquid, and gaseous contaminants produced by man.

The "heat island" is one of the chief features of the urban climate. It is characterized by higher air temperatures than rural areas. This effect has been investigated in many experimental works, and several numerical models have been proposed for its study.

Results of previous investigations (SMIC, 1971; Landsberg, 1970; Berlyand and Kondrat'ev, 1972) have demonstrated that a heat island is usually of complex structure, each city block of an urban development being a source of heat for surrounding undeveloped sections. The mean air temperature in a large city is often 1–2° greater than the temperature of surrounding regions, although the temperature difference may reach 6–8° at night with a light wind. This difference usually decreases with strong winds.

Landsberg and Maisel (1972) carried out interesting observations characterizing the growth of a heat island as a city develops. In 1967, in the USA near Washington, D.C., the construction of the city of Columbia was begun. In 1968, when only a small fraction of the city's buildings had been completed, the air temperature was 0.5° higher in its center at night than it was in the suburbs. In 1970, when the city was significantly developed, the temperature difference reached 4.5° in its center, and 2° over a larger part of its area. The temperature increased less in the daytime when the intensity of air movement was greater. At the same time, as the temperature increased over the area of Columbia, the relative humidity decreased by several percent, this decrease apparently being caused by both the increased temperature and decreased evaporation throughout the city.

It is obvious that a heat island appears mainly as a result of the effect of the first three factors enumerated above, which govern urban climatic conditions. The relative role of each of these factors may vary strongly in different cities and seasons.

This conclusion is confirmed by the results of theoretical studies of the urban thermal regime. For example, Atwater (1977) has calculated temperature variations in cities for the four climatic zones (tropics, desert, middle latitudes, and tundra), taking into account the thermal regime effect of anthropogenic heat income, the state of the earth's surface, and the amount of aerosol in the atmosphere. Atwater in his calculation has used the heat balance equation of the earth's surface. He has established that the greatest temperature increase is observed in a city situated in tundra, whereas in the tropics and in deserts the temperature changes only slightly. The heat island effect becomes most apparent during the night. The calculations show that an increase in air temperature in the cities depends essentially on anthropogenic heat income, whereas variations in the atmospheric aerosol content affect the thermal regime slightly.

As observations show, in heat islands not only relative but also absolute air humidity usually decreases due to reduced evaporation in built-up regions.

Additional heating of the air over cities creates local circulation systems resembling sea breezes and also amplifies ascending convective motions over the cities. In addition, increased roughness of the surface brings about a noticeable decrease in the wind speed in cities as compared with the suburbs.

Among all climatic features, of most importance is air pollution by different contaminants, which reaches high levels in many cities. Industrial enterprises, heating systems, and traffic pollute the atmosphere.

An increased aerosol concentration over cities decreases the solar radiation incoming to the earth's surface. According to Landsberg (1970), direct solar radiation in large cities is frequently decreased by about 15%, ultraviolet radiation on the average by 30% (it may entirely vanish during winter months), and the period of solar irradiation by 5–15%. The thin atmospheric lower layer containing the greatest amount of aerosol particles is of major importance in attenuating solar radiation in cities. Horizontal visibility is usually drastically reduced within this layer, sometimes to 10–20% of its value in rural regions.

Due to high aerosol concentration in urban air, fog, particularly smog (a type of stable fog whose droplets contain a significant amount of the contaminants that pollute the atmosphere) becomes more frequent. In some cities (Los Angeles is a particularly well-known example), where local conditions promote the attentuation of atmospheric circulation, smog may remain for many days, greatly harming the health of the population.

Urban fog plays an important role in decreasing solar radiation and reducing the range of visibility in cities.

An increased amount of condensation nuclei in the air over cities and strengthening of the ascending air motions result in increased cloudiness and precipitation. Available data indicate that a week-long cycle in precipitation amount exists in some industrial centers, i.e., precipitation decreases on weekends when industrial enterprises stop operating.

Due to some increase in precipitation and considerable decrease in evaporation, runoff grows. In many cities increased runoff is facilitated by surface water sewerage systems. Snow removal and its transportation outside the city are often used in cities with great amounts of solid precipitation; this accelerates temperature elevation in the city during spring in comparison with the suburbs.

Urban climate may be significantly improved by rational siting of residences and plants, by expanding green areas, and by reducing air pollution.

There are many examples illustrating how a change in heating system (conversion from solid fuel to gas or electric energy) drastically reduces urban air pollution and fluctuations of related climatic parameters. Of no less importance is the removal of industrial enterprises from the city, the use of an effective method for purification of air entering the atmosphere from chimney stacks and other sources of pollution, the laying out of parks within a city, and the planting of trees and shrubs along the streets and on vacant land.

At the contemporary level of progress in industry, energetics, and transport, urban construction cannot be carried out without a detailed study of man's impact on the meteorological regime. The calculation of expected changes in local climate is necessary for proper city planning to avoid the development of unfavorable climatic conditions.

It is worth mentioning that, as urbanization progresses in regions with high population density, some features of urban climate spread over vast areas reaching hundreds of square kilometers. Under these conditions, there is a cumulative climatic effect of numerous closely situated populated areas acting as sources of air heating and pollution. This situation occurs in a number of regions in the USA, Japan, and western Europe, where local climatic changes cover evergrowing areas.

5.2 CHANGES IN ATMOSPHERIC COMPOSITION

Carbon dioxide concentration growth

As indicated in Chapter 1, many years ago Callendar (1938) assumed that the burning of coal, oil, and other kinds of fossil fuel resulted in increased CO_2 mass in the atmosphere.

Although measurements of atmospheric CO_2 concentration have been carried out since the 1850s, it was difficult to verify this assumption because of the inadequate precision of the instruments used in the detection of comparatively small variations in CO_2 concentration.

Since the late 1950s the CO_2 concentration has been systematically observed with the help of precise instruments at two stations — Mauna Loa (Hawaii) and the South Pole. Observation results obtained at Mauna Loa are presented in Fig. 5.2 (Keeling and Bacastow, 1977). Curve 1 shows variations in monthly means of CO_2 concentration in the period 1958–1976. From this figure it is seen that photosynthesis exerts a noticeable influence on the annual cycle of CO_2 concentration, decreasing its values by 2% during the periods of increased productivity of autotrophic plants. Curve 2 depicts the CO_2 concentration values obtained by excluding the annual cycle effects by means of averaging.

Data in Fig. 5.2 show that CO_2 concentration grows from year to year. A similar conclusion follows from the data of observations at the South Pole station and several other stations that began operating later.

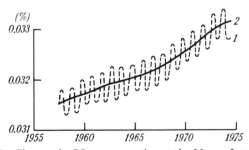

Fig. 5.2 Changes in CO_2 concentration at the Mauna Loa station.

At present there are about 12 stations at which observations of CO_2 concentration are carried out. Most of them have begun operating only recently. Longer series of observations are available for the Point Barrow (Alaska) station and for Europe, where measurements were carried out by Scandinavian aircraft. These observations give similar results, showing a year-to-year increase in CO_2 concentration.

In the measurements of CO_2 concentration at Mauna Loa and the South Pole it was established that during 1959–1968 the concentration of CO_2 increased on the average of 0.74 ppm yr^{-1}. This magnitude corresponds to approximately 0.25% of the amount of CO_2 contained in the atmosphere. By 1974 the increase in CO_2 reached 1 ppm (or 0.34%) per annum (Olson *et al.*, 1978).

Figure 5.2 shows an increase in the rate of accumulation of CO_2 in the atmosphere. As seen, during the second part of the period in question, the rate of CO_2 concentration growth was greater than in the first part.

Since these observations of CO_2 concentration were carried out at stations situated at great distances from each other, there is no doubt that they correctly indicate a present-day global tendency in variations of atmospheric CO_2 concentration. Observation data show that over the period 1958–1978 the mass of atmospheric CO_2 has increased by approximately 5%.

The question of how the CO_2 mass has increased over the last 100–130 yr, i.e., for the period of intensive consumption of fossil fuel, is more difficult to answer. Using observational data for the 19th century (whose accuracy is limited), Machta (1978) concluded that between 1858 and 1958 the mass of CO_2 increased by 3 to 15%. Thus, up to now, the total CO_2 concentration increase is 8–20%. To verify this estimate, we can use the models of atmospheric CO_2 balance that are discussed more thoroughly in the next chapter. In a number of studies it has been established that from the middle of the past century to the late 1970s the CO_2 concentration has grown from 0.029% to 0.033%, i.e., by approximately 14%, taking into account only one factor—the burning of fossil fuel. Recent studies of CO_2 balance that consider the effect of varied vegetation mass lead to the conclusion that the increase in CO_2 concentration for the time period in question might have been somewhat larger.

The balance of carbon dioxide

In the present epoch the balance of CO_2 has a much greater dependence on anthropogenic factors than on natural conditions.

From data given in Chapter 2 it follows that among the components of the natural CO_2 cycle, the largest is the expenditure of CO_2 in photosynthesis, amounting to 300×10^{15} g yr^{-1}. This value, however, is almost exactly compensated for by the amount of CO_2 formed as organic matter oxidizes. The difference between the income and outgo of atmospheric CO_2 in the biotic cycle, according to data on accumulating organic carbon during the Pliocene, amounts to 0.05×10^{15} g yr^{-1}. The total outgo of CO_2 in the formation of carbon-containing sedimentary rocks during this period of time is 0.09×10^{15} g yr^{-1}. The total value, 0.14×10^{15} g yr^{-1}, should correspond to the income of CO_2 into the atmosphere during volcanic eruptions, from hot springs, etc. It is very difficult to determine the value of this income. Bowen (1966) believes that for the present epoch it corresponds to an amount of carbon equal to 0.04×10^{15} g yr^{-1}, i.e., 0.14×10^{15} g yr^{-1} of CO_2. The agreement between the above values of income and outgo of atmospheric CO_2 with their limited accuracy should be considered to a large extent random. Since we are interested only in the order of magnitude of the corresponding CO_2 balance component, one can consider it to be about 0.1×10^{15} g yr^{-1}. This value should be compared with the income of CO_2 to the atmosphere due to man's economic activities.

The burning of coal, petroleum, and other kinds of fossil fuel, which supplies at present more than 97% of the total energy consumed by man, is the major form of man's impact on the atmospheric CO_2 balance. The expenditure of fossil fuel over the last 25 yr grew by 4% per annum. In the late 1970s it reached a value corresponding to an income to the atmosphere of about 5×10^{15} g of carbon or 18×10^{15} g of CO_2 yearly. Taking account of the effect of other kinds of industry on the balance of CO_2 changes this value slightly. For example, the manufacture of cement releases a quantity of CO_2 that increases its amount by only 2% (Keeling and Bacastow, 1977; Olson et al., 1978).

Less clear is the question of the amount of carbon entering the atmosphere due to man's impact on vegetation cover. Until recently this component of the CO_2 balance has not been taken into account, though Hutchinson in 1954 suggested that the income of carbon into the atmosphere from biotic sources may be compared with that from fossil fuel (Woodwell, 1978). Only in the last few years has this question attracted attention. Bolin (1977), using data on annual consumption of wood, came to the conclusion that, because of the cutting of forests and destruction of soil humus, about 1×10^{15} g yr^{-1} of carbon enter the atmosphere at present. Woodwell and Houghton have proposed a far greater estimate of this value that, in accordance with a recent publication by Woodwell (1978), amounts to 8×10^{15} g yr^{-1} of carbon.

In papers by several authors from Oak Ridge National Laboratory (Baes et al., 1976; Olson et al., 1978), it is believed that as a result of the felling of tropical woods, about 1×10^{15} g yr^{-1} of carbon are released and about 2×10^{15} g of carbon are produced in the course of the decomposition of previously created sources of organic matter. The authors of reviews of atmospheric carbon balance compare a relatively high accuracy in estimating the carbon income to the atmosphere resulting from industrial activity (it is assumed that the error of this estimate does not exceed 5%) to a low reliability in calculating the carbon income due to the impact on vegetation cover and soils. These calculations are based on a limited number of data; the lack of data is made up for by more or less probable hypotheses. Nevertheless, as seen from recent studies, this component of the atmospheric carbon balance may be highly significant.

Comparing the estimates of the income of anthropogenic CO_2 to the atmosphere, which are of the order of $n \times 10^{16}$ g yr^{-1}, with those of the income from natural sources, which are 10^{14} g yr^{-1}, it is easy to see that man's economic activity is the controling factor in the contemporary balance of atmospheric CO_2.

To study variations in the CO_2 balance over the last 100–150 yr we must have data on the amount of fossil fuel used and on the decrease in the amount of carbon accumulated in living organisms and the products of their vital activity referring to this time. It is assumed that from 1860 to 1973 128×10^{15} g, and from 1860 to 1959 81×10^{15} g, of carbon from fossil fuels entered the atmosphere (Keeling and Bakastow, 1977).

Bolin (1977) believes that, due to a decrease in biomass from the early 19th century to our time, about 70×10^{15} g of carbon have entered the atmosphere. Revelle and Munk (1977) concluded that for 1860–1970 this amount is 72×10^{15} g. These two estimates are based on very incomplete evidence.

To study the balance of atmospheric carbon, Stuiver (1978) and Wilson (1978) have utilized the method of isotopic analysis. Stuiver estimated the amount of isotope ^{14}C in tree rings. The ^{14}C in the atmosphere is produced by the action of cosmic rays and is comparatively unstable (it has a half-life of 5,700 yr). Therefore, ^{14}C has long ago disappeared from fossil fuels, which were formed millions of years ago. The combustion of fossil fuels releases CO_2 to the atmosphere that lowers the atmospheric ^{14}C level (the Suess effect).

In addition, fossil fuels as well as plants contain a smaller amount of the stable ^{13}C isotope than the atmosphere because, during photosynthesis, plants discriminate against ^{13}C in favor of the lighter ^{12}C.

Thus data on the ^{13}C and ^{14}C isotope content of tree rings of different ages allow us to estimate the contribution of fossil fuels to the total atmospheric CO_2 mass (which reduces the atmospheric ^{14}C and ^{13}C levels) and of the oxidation of plant material (which reduces the atmospheric ^{13}C level).

Using this method Stuiver has obtained a somewhat unexpected result: between 1850 and 1950, about 120×10^{15} g of carbon were injected into the atmosphere from the biota. This value is twice the release of carbon from the burning of fossil fuels (60×10^{15} g) over the same period. Stuiver has established that the fraction of carbon from fossil fuels rapidly increased and the fraction from the biota decreased. As a consequence, by 1970 the cumulative amount of carbon produced from each source appeared to be approximately the same.

Applying the method of isotopic analysis, Stuiver has estimated the amount of CO_2 absorbed by the ocean during 1850–1950 to be approximately half of the income from the burning of fossil fuels and from biota. This results in an atmospheric CO_2 content increase of about 90×10^{15} g of carbon.

Such a large injection of carbon into the atmosphere is possible with a comparatively low atmospheric CO_2 concentration at the beginning of the period in question. According to Stuiver this concentration was 0.0268%. This means that the modern concentration exceeds the concentration at the middle of the nineteenth century by somewhat more than 20%. This value is close to the upper limit of possible concentration change mentioned by Machta. Wilson has obtained similar results. Using the same approach, he found that the CO_2 concentration of the earth's atmosphere increased in the second half of the nineteenth century by about 10%. He concluded that this was due to the expansion of agricultural lands and deforestation and that at the middle of the nineteenth century the concentration of CO_2 amounted to 0.027%. This value almost exactly equals that obtained by Stuiver. Using data obtained by Stuiver and Wilson in studying the CO_2 balance is rather difficult because the accuracy of the data is not very high.

The question of the outgo of anthropogenic CO_2 appears to be complicated enough. Undoubtedly the ocean plays an important role in absorbing excess CO_2.

The total amount of dissolved CO_2 greatly exceeds the atmospheric CO_2 content. However, as the results of studying turbulent exchange in the ocean showed, trace gases in the atmosphere easily penetrate by this means only into the upper mixed layer of ocean water—about 100 m deep. Below this layer the diffusion processes are significantly weakened, preventing the excess CO_2 from being transferred into deeper layers of ocean water. Since the upper well-mixed layer can absorb only a limited amount of CO_2, the potential of the ocean to absorb a greater portion of anthropogenic CO_2 is questionable. Cold waters in high latitudes probably play an important role in the absorption of CO_2 by the ocean, carrying away the dissolved CO_2 into deeper layers as they sink.

A certain amount of CO_2 can be assimilated by autotrophic plants whose biomass, other things being equal, depends on CO_2 concentration, increasing as it rises. The growth of the biomass of photosynthesizing plants must be followed by an increase in the mass of related organisms and the products of their vital activity.

Summarizing the results of studying the present CO_2 cycle, the authors of the report "Energy and Climate" (National Academy of Sciences, 1977) conclude that over the last 110 yr the amount of carbon in the atmosphere has increased by 72×10^{15} to 83×10^{15} g. Over the same time, the amount of CO_2 released by burning fossil fuels and manufacturing cement contained 127×10^{15} g of carbon and the amount resulting from cutting forests and cultivating the soil was about 70×10^{15} g. It is believed that 40% of anthropogenic CO_2 has remained in the air, 20% has gone into the oceans, and 40% has gone to the biosphere (as an increase in the mass of continental organisms and the products of their vital activity). The increase in the atmospheric CO_2 concentration that occurred in the last century must have changed the thermal regime. This is discussed below.

Other gases

The atmospheric thermal regime can be affected in certain ways by other gases entering the atmosphere as a result of man's economic activities. Fluorocarbons (Freons), widely used as refrigerants and as solvents in the manufacture of various paints, can be important among these other gases. When the paints dry, the Freons evaporate into the atmosphere where they are able to survive for many years because their lifetime, which depends on the rate of photodissociation in the upper atmospheric layers, amounts to tens of years.

Freons are transparent to other wavelengths but absorb intensely infrared radiation with wavelengths of $8-12$ μm, i.e., in the "transparency window" where almost no longwave emission is absorbed by water vapor. Therefore, they can considerably affect the surface air temperature. Ramanathan (1976) has shown that a concentration of Freons amounting to 2×10^{-9} of the atmospheric volume can increase the temperature near the earth's surface by almost $1°$. Wang et al. (1976) have obtained a somewhat lower value of the temperature increase, amounting to $0.3-0.4°$. Since at present the atmospheric concentration of Freons is

$0.1–0.2 \times 10^{-9}$ of the volume of the atmosphere, their effect on the thermal regime is not great.

Among other gases whose concentration variations may have exerted effects on the thermal regime are nitrogen compounds, particularly N_2O. Unlike Freons, which appear in the atmosphere only as a result of economic activities, N_2O is always present, amounting to about 0.3×10^{-6} of the atmospheric volume. Doubling this quantity results in a $0.5°$ temperature increase near the earth's surface (Yung *et al.*, 1976). Since N_2O is formed when fossil fuels are burned and nitrogen fertilizers are used, its atmospheric amount may depend on man's economic activities. There are some other gases released from industrial chimney stacks that exert some influence on the thermal regime of the atmosphere. Among them are CO and CH_4 (methane). An increase in their mass due to growing consumption of fossil fuels could lead to an elevation of the tropospheric ozone concentration (Hameed *et al.*, 1979) that would result in intensification of the greenhouse effect.

It should be indicated that the probability of significant climatic modification due to the action of growing concentrations of Freons and other gases representing minor atmospheric contaminants is considerably less than that for CO_2. This is true because the mass of these gases is comparatively small and its changes may be controlled by means of rational planning without damaging major branches of the economy. In contrast, preventing further growth of CO_2 concentration in the atmosphere is extremely difficult. This will be discussed below in more detail.

It is doubtful whether it is possible to stop the increase in concentration of minor contaminant gases in the atmosphere when their mass depends on the amount of fuel burned. It is thought that this growth might cause some temperature change at the earth's surface in the direction of warming.

Keeping in mind the necessity of controling the concentration of minor contaminant gases in the atmosphere that affect climate, it might be supposed that their influence would be less than that of CO_2 concentration growth.

Atmospheric aerosol

As mentioned above, due to man's economic activities, a large number of particles enter the atmosphere, increasing the concentration of atmospheric aerosol.

Available estimates (SMIC, 1971) indicate that about 200–400 million tons of anthropogenic aerosol is now ejected yearly into the atmosphere, comprising 10–20% of the total amount of aerosol particles that enter the atmosphere. Only a small portion of this amount is ejected into the atmosphere in the form of solid or liquid particles; the chief sources are the gaseous compounds produced by man, such as sulfur dioxide and nitric oxides, from which aerosol particles are formed as a result of various chemical reactions.

Undoubtedly, man's economic activities cause an increase in aerosol concentration in many cities and industrial regions and in agricultural areas where cultivation of the soil intensifies its erosion. The extent to which anthropogenic aerosol spreads over vast areas and penetrates into high atmospheric layers is a much more complex question.

The stratospheric aerosol that noticeably influences climatic conditions is assumed to be natural. The anthropogenic sulfur dioxide produced in particular by the burning of sulfur-containing coal and oil seems not to enter the stratosphere, because current direct observations of stratospheric aerosol record very low concentrations of it for some years. Thus the stratospheric aerosol is chiefly formed by natural processes, in particular those associated with volcanic activities.

Davitaya (1965) was the first to assume that the anthropogenic aerosol spreads over vast areas. He used data on the vertical distribution of dust concentration in the snow cover of Caucasian glaciers. These data indicated that the amount of dust per unit volume in the upper snow layers significantly exceeded that in the deeper layers. Davitaya believed that this difference corresponded to an increase in dust concentration in the atmosphere that occurred during the last decades.

MacCormic and Ludwig (1967) have presented data indicating a decrease in the transparency of cloudless atmosphere over the last several decades, caused apparently by an increased concentration of atmospheric aerosol.

In many studies (MacCormic and Ludwig, 1967; Bryson, 1968; Budyko, 1969a; SCEP, 1970; SMIC, 1971; and elsewhere) it has been proposed that a gradual decrease in direct radiation observed at a group of actinometric stations beginning in the 1940s can be attributed to the increasing mass of anthropogenic aerosol in the atmosphere.

Dyer (1974) and Elsasser (1975) have concluded that during the last decades the amount of aerosol over vast areas has not grown. There are some data contradicting this opinion. For example, N.A. Efimova has processed data from observations of direct radiation under a clear sky for a number of actinometric stations situated at great distances from cities and industrial regions and concluded that in the 1970s the atmospheric transparency was less than that in the 1950s. Kellogg (1975) has constructed, using available data, a world map of anthropogenic aerosol distribution showing that increased aerosol concentrations are observed over Europe, most of Asia and North America, and the northern Atlantic Ocean.

Pivovarova (1977a) has studied the question of the varied mass of atmospheric aerosol. She used radiation data of observations for periods of 40 to 70 yr obtained at 13 stations in the USSR and the USA. Pivovarova established that, over the last decades at these stations, direct solar radiation under a clear sky has decreased. This process is very well observed, particularly over the USSR European Territory and Kazakhstan. Less radiation variation takes place in Siberia, the Far East, and the south of Central Asia. A decrease in radiation was fixed at both urban and rural stations. Therefore, we may think that the aerosol mass has increased, not only in cities but also over vast areas covering a considerable part of the continents in the mid-latitudes in the Northern Hemisphere. This conclusion agrees well with that of Kellogg.

The data of Pivovarova possibly reflect variations both in manmade and in natural aerosol. It is believed that the question of changing the amount of anthropogenic aerosol over vast areas requires further investigation.

The particles of aerosol in the stratosphere act like a screen, changing to a lesser or greater extent the meteorological solar constant. Absorption of radiation by these particles can cause local heating of the stratosphere, as occurred, in particular, after the eruption of the Mount Agung volcano (Newéll, 1971), although this heating scarcely affected the thermal regime at the earth's surface because of insignificant air density in the stratosphere and weak heat exchange between the stratosphere and the troposphere. Thus, as has been demonstrated in a number of studies, an increase in aerosol concentration in the stratosphere invariably leads to a temperature drop at the earth's surface.

The tropospheric aerosol particles exert a more complex effect on the thermal regime. They attenuate the flux of shortwave radiation reaching the earth's surface as a result of backscattering and absorption of emission by the aerosol particles. While the first process increases the albedo of the earth–atmosphere system, the second may decrease it.

Vertical redistribution of absorbed radiation within the troposphere, where the vertical heat exchange is intensive, influences but slightly the mean tropospheric temperature or the surface air temperature. Therefore, the influence of tropospheric aerosol on the thermal regime is basically determined by the dependence of the earth–atmosphere albedo on aerosol concentration. An increase in albedo leads to a temperature drop at the earth's surface, while its decrease leads to a temperature rise.

Obviously, a change in the albedo of the earth–atmosphere system due to aerosol will depend on the albedo of the earth's surface. The lower this albedo is, the more likely it is that atmospheric aerosol will increase the system's albedo. For a high albedo of the earth's surface (snow or ice) the probability of decreasing the earth–atmosphere system's albedo due to aerosol effect grows.

Numerical models have been constructed in several studies to describe the influence of aerosol on radiation and the thermal regime of the atmosphere (MacCormic and Ludwig, 1967; Charlson and Pilat, 1969; Atwater, 1970; Barrett, 1971; Yamamoto and Tanaka, 1972; Rasool and Schneider, 1971; Mitchell, 1971; Ensor *et al.*, 1972; Newman and Cohen, 1972; Kondrat'ev *et al.*, 1973; Department of Transportation, 1975; Pollack *et al.*, 1975, 1976). Some of these studies took into account only the influence of aerosol on the backscattering of shortwave radiation, which results in increased albedo of the earth–atmosphere system and consequently in decreased air temperature near the earth's surface.

Other studies have considered fluctuations in the atmospheric radiation and thermal regime caused by both backscattering and absorption by aerosol particles. The second of these mechanisms can decrease the albedo of the earth–atmosphere system under certain conditions, contributing to a temperature increase near the earth's surface.

Quantitative criteria that depend chiefly on the albedo of the earth's surface and determine the sign of the temperature variations due to the action of aerosols have been proposed in a number of studies. The values of these criteria found by different authors are noticeably different. This is explained both by the particular features of the models used and by the fact that the values of the empirical coefficients characterizing backscattering and absorption of radiation by aerosol particles do not coincide.

In several studies it has been assumed that the anthropogenic aerosol exerted an effect on climate that influenced the formation of clouds and changed their top albedo (SMIC, 1971; National Academy of Sciences, 1977). This assumption is not satisfactorily proved and should be studied.

In summary, it should be noted that although the question of the global climatic effect of anthropogenic aerosol remains in some apsects unsolved, the probability exists that this effect is comparatively small. The reason for this is, as shown below, that the secular trend of mean air temperature could be satisfactorily described by considering the effect on the temperature of CO_2 concentration and fluctuations in atmospheric transparency due to the varied mass of stratospheric natural aerosol.

It is worthwhile to raise the question of what climatic effect the anthropogenic aerosol could exert if it spread in the stratosphere, where the lifetime of particles far exceeds that in the troposphere.

As noted above, anthropogenic sulfurous gases produced by burning fuel at the earth's surface seem not to enter the stratosphere. However, these gases can be produced by aircraft operating in the stratosphere.

The question of the climatic effect of SST aircraft operations was studied in the CIAP project during the first half of the 1970s in the USA (Department of Transportation, 1975).

To assess variations in the physical state of the atmosphere caused by stratospheric flights, the content of nitrogen oxides and sulfur in the discharge of turbojet engines was calculated in the above study, taking into account the numbers of subsonic and supersonic airplanes.

Of the various ways in which SSTs can pollute the stratosphere, the discharge of SO_2 can exert the greatest effect on climate. Gaseous SO_2 entering the stratosphere is oxidized and reacts with water vapor, producing sulfuric acid droplets. These absorb some water vapor from the air and, as a concentrated solution of sulfuric acid, remain in the stratosphere for a long time—up to several years. Some of the droplets, after reacting with ammonia, are converted into ammonium salts. However, a greater portion do not change their chemical composition before they fall into the troposphere because of increasing droplet sizes due to coagulation, gravity sedimentation, and the action of air motions.

Usually the size of a sulfate particle is several tenths of a micron. As noted above, particles of this size strongly affect shortwave radiation, attenuating its flux by absorption and backscattering.

Therefore, variations in stratospheric aerosol mass can change the temperature of the atmospheric lower layer and other climate elements.

For quantitative estimation of the effects of stratospheric aircraft on climate, the dependence of the optical thickness of the stratosphere on the mass of aerosol particles in the stratosphere has been calculated. Under natural conditions, the optical thickness of the stratosphere is about 0.02. This means there is a 2% attenuation of the radiation flux passing through the stratosphere. The dependence between variations in the optical thickness of the stratosphere $\Delta\tau$ and the sulfate aerosol concentration ΔS in micrograms per cubic meter can be expressed in the form

$$\Delta\tau = 0.038 \ \Delta S. \tag{5.6}$$

Taking into account the mean lifetime of particles, their density, the mean concentration of H_2SO_4 in droplets, and the fact that the bulk of the sulfate aerosol is within the 10-km layer, the following formula can be obtained:

$$\Delta\tau = 5.1 \times 10^{-11} \ E, \tag{5.7}$$

where E is the mass of SO_2 (in kilograms per year) discharged over one hemisphere by 20-km-level flights. The coefficient in this formula is reduced by half for 17-km-level flights.

In addition to the effect of stratospheric aerosol on shortwave radiation, an increase in its mass also decreases the flux of longwave emission, thus strengthening the greenhouse effect. The calculations show that the decrease in shortwave radiation is four times greater than that in longwave emission. Thus the latter effect can be neglected in studying variations in the radiation regime.

The climatic models establishing the relationship between the income of solar radiation and air temperature near the earth's surface should be used in assessing the climatic effect of variations in radiation income to the troposphere. Because of the inertia of the ocean–atmosphere–polar-ice system this dependence is different for different periods of time during which temperature varies. Assuming that the process studied occurs over 10 yr, the authors of this investigation have obtained for these conditions the relationship between the relative change in the income of total radiation $\delta\sigma/\sigma$ and the change δT in the mean annual planetary air temperature at the earth's surface in the form (Budyko, 1968a)

$$\delta T = 150 \, \delta\sigma/\sigma. \tag{5.8}$$

The proportionality coefficient in this formula increases with increasing period of time. From the formulas of atmospheric optics, the relationship between the relative variations in total radiation and stratospheric optical density for this case was found to be

$$\delta\sigma/\sigma = -0.186 \, \Delta\tau. \tag{5.9}$$

Using Eqs. (5.7)–(5.9) we can find variations in the mean air temperature δT for the cases of different discharges of SO_2 into the stratosphere.

Assuming that in the year 2000 high-flying aircraft will produce 10^5 tons yr^{-1} of discharges, we can find that the mean temperature in this case will be lowered by 0.16°.

In addition, the atmospheric radiation regime may be affected by a decrease in ozone concentration and an increase in the amount of water vapor and nitrogen oxides discharged by aircraft in the stratosphere.

Because of decreased ozone concentration the atmospheric absorption of solar radiation decreases and the temperature drops. This is compensated for to some extent by an increase in absorption of radiation due to the growth of NO_2 concentration. Taking this into account we find that with a 10% decrease in total ozone concentration the temperature of the surface air layer is lowered by approximately 0.1°. Such a variation in ozone mass can occur only with a considerable mass of discharges from stratospheric aircraft engines. Under these conditions an additional quantity of water vapor enters the stratosphere, increasing its humidity by 20–30%. This intensifies the absorption of outgoing longwave emission, thus leading to an air temperature rise near the earth's surface of 0.1–0.2°. Thus the total effect on the thermal regime of changing ozone and water vapor concentrations may prove to be insignificant, even with large amounts of aircraft discharges.

From CIAP data we may conclude that at present the climate is only slightly affected by stratospheric aircraft and that in the next 10 to 20 yr the effects are not likely to become significant.

5.3 CHANGES IN GLOBAL CLIMATE

Effects on atmospheric composition

As noted in the previous section, man's impact on the chemical composition of the atmosphere must have resulted in varied global climatic conditions due to the growth of CO_2 concentration. The effect of man-made aerosol on the global climate seems to be less significant although it changes somewhat climatic conditions over many populated areas.

Thus the main question of the problem under consideration is associated with studying the effect of CO_2 concentration growth on climate.

As noted before, at present there are two estimates regarding the CO_2 concentration growth that occurred from the mid-1850s to the late 1970s. Taking into account the amount of fossil fuel consumed, it was concluded that during this period the CO_2 concentration has increased from 0.029 to 0.033%, i.e., by 14%. Stuiver (1978) and Wilson (1978) concluded that over this period the CO_2 concentration has grown from 0.027 to 0.033%, i.e., by 22%. These results could be in agreement if a conclusion of Stuiver and Wilson were taken into account: that the CO_2 concentration growth was greatly affected by the destruction of forests that occurred, as they believed, chiefly in the second half of the 19th century. The bulk of the fossil fuels consumed since the mid-1850s was burned during the 20th century. Therefore, the CO_2 concentration seems to have been increased by 10–12% in the second half of the 19th century chiefly because of a decreased mass of plants and products of plant destruction and increased by 10–12% in the 20th century because of the burning of fossil fuels. The latter value is more reliable because the precision of the isotope method used by Stuiver and Wilson is rather limited.

The values of CO_2 concentration growth given here are not small. This growth must have exerted a considerable effect on the climate and the productivity of vegetation cover (the latter will be discussed in Chapter

6). Here we give the estimation for the variations in mean air temperature near the earth's surface that must have occurred under the influence of increasing CO_2 concentration.

In Fig. 5.3 are presented data showing variations in the mean air temperature near the earth's surface in the Northern Hemisphere computed for 5-yr running periods (curve 2) and variations in this temperature due to CO_2 concentration growth (curve 3) for the last 90 yr. This calculation (which is mentioned in Chapter 3 and discussed in more detail in Chapter 6) is carried out with the assumption that doubling the CO_2 concentration while retaining continental polar ice leads to a 3° increase in mean surface air temperature. In addition, the nonlinear dependence of the temperature increase on CO_2 concentration growth is taken into account.

As seen from Fig. 5.3, although actual temperature variations differ from calculated results, observational data confirm the tendency toward warming. This tendency is easy to see in a comparison of the temperatures for the first and the second halves of the period in question. For example, before 1920 the temperature did not reach the mean value (normal) for the entire period. After that year it exceeded the normal for almost the whole interval. Curves 2 and 3 do not coincide because of the effect of variations in atmospheric transparency on the temperature. We have established that as the transparency decreases, the temperature decreases and that increased transparency causes increased temperature. This is confirmed by comparing data on relative anomalies of solar radiation under a clear sky that describe the atmospheric transparency, (curve 1) and the secular trend of air temperature (curve 2).

Figure 5.3 shows that the growth in CO_2 concentration has resulted in a

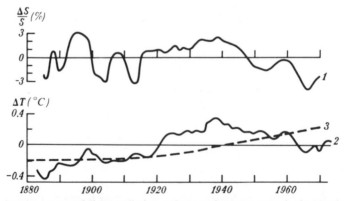

Fig. 5.3 Secular course of direct radiation and mean air temperature in the Northern Hemisphere: curve 1, direct radiation; curve 2, mean air temperature; curve 3, temperature variation due to CO_2 mass growth.

0.5° temperature rise near the earth's surface, beginning at the end of the 19th century. Although this value is sufficiently great, it is nevertheless considerably disguised by short-term climatic changes caused by variations in atmospheric transparency.

The question of how the mean air temperature varied in the 19th century should be further studied. If during the second half of the 19th century the CO_2 concentration increased by 10–12%, the mean global temperature should have increased by 0.5°. As a result, the total variation, due to man's impact, in the mean temperature from the mid-1850s until now is about 1°.

This conclusion can hardly be confirmed by observational data because a more or less complete meteorological network on the continents was established only at the end of the 19th century. Observations of the state of Arctic Sea ice, which is an indicator of the thermal regime, were carried out only in a few regions at high latitudes in the 19th century. Available data show that a certain temperature rise did take place in the second half of the 19th century, although it seemed to be less than the indicated value.

It is likely that the warming caused by the growth of CO_2 concentration during the second half of the 19th century was considerably compensated for by the thermal regime effect of an increase in global albedo of the earth–atmosphere system due to deforestation, which will be discussed below in more detail.

It is easier to estimate quantitatively the influence of variations in CO_2 concentration on the air temperature during the past 90 yr. As seen from Fig. 5.3, at this time the air temperature in the Northern Hemisphere, averaged over 5-yr running periods, depended on the growth of CO_2 concentration and on fluctuations in atmospheric transparency. Curve 1 in Fig. 5.3 presents values of direct radiation anomalies under a clear sky as obtained by Pivovarova (1977b). These values are averaged over 5-yr running periods.

It should be noted that curves of secular temperature trends are constructed from observational data from hundreds of meteorological stations situated on the continents and from ships' observations. Thus, the temperature data cover the extratropical zone in the Northern Hemisphere more or less evenly.

By contrast, data on the secular trend of direct radiation are obtained from a group of actinometric stations situated in the mid-latitudes of Europe, Asia, and North America. At the beginning of the period in question data from a few stations were used; at the end of the period there were 13 stations.

The analysis of data by Pivovarova shows that it is sufficient to have data from only ten actinometric stations, situated in different regions of

the Northern Hemisphere, in order to reveal the major features of variations in the mean value of direct radiation. Therefore, radiation data given
in Fig. 5.3 can be considered as representative for a greater part of the
period in question. Although the available estimates of radiation anomalies for the beginning of this period are not adequate, they do in this case
give useful information on the conditions of atmospheric transparency.

Comparison of curves 1 and 2 is of great interest. The chief difference
between them is the tendency to rising temperature observed over a
greater part of the period in question. There is no such tendency in the
curve of the secular trend of direct radiation, which is characterized by
fluctuations both above and below a stable mean value, with some reduction in radiation during the last decades.

In addition, there is a close similarity between the main features of the
secular trend of direct radiation and the temperature.

It is seen that the chief maxima and minima of curve 2 correspond to
similar variations in curve 1, the temperature variations lagging somewhat
behind those in radiation and being smoother.

Of great importance is the relationship between radiation fluctuations
and a long-term warming trend, which began in the 1920s. In 1914–1915, a
sharp increase in radiation occurred, following which a long-term positive
anomaly in radiation was observed. In 1918–1922, due to increased radiation, the air temperature rose. Then in 1931–1934, radiation increased
again, causing a temperature elevation in 1935–1938. The 1953 radiation
minimum resulted in the 1956–1957 temperature minimum. The radiation
peak in 1959 was followed by the maximum in temperature in 1960. The
deep minimum in radiation in 1966 led to an air temperature minimum in
1967. Thus, in most cases considerable increases and decreases in radiation were followed by temperature fluctuations that lagged behind the radiation variations by a period of one to five years (more often by three
years).

Observational data show that variations in radiation income to the troposphere depend chiefly on the mass of stratospheric aerosol, which
changes slightly in space within the stratosphere over each of the hemispheres because of intensive horizontal mixing. As a result, for estimating
the average conditions of stratospheric transparency, data of actinometric
observations (or observations of aerosol concentration) at a small number
of stations are sufficient.

The influence of atmospheric transparency variations and changes in
CO_2 concentration on air temperature can be studied using the aforementioned dependences. As seen, the sign of the difference between the temperature anomalies depicted by curves 2 and 3 in most cases coincides
with that of the direct radiation anomalies. This demonstrates the deter-

mining character of the effect of radiation fluctuation on less extended air temperature variations in the Northern Hemisphere.

It is interesting to compare the relative values of temperature anomalies caused by radiation fluctuations with those of the radiation anomalies. Figure 5.3 shows that the ratio of temperature anomaly due to radiation fluctuation to radiation anomaly increases with the period of time during which the sign of the corresponding anomalies remains the same. From 1920 to 1945 the longest positive radiation anomaly was observed, its mean value being $+1.3\%$. From calculations by formulas of atmospheric optics it follows that this corresponds to a 0.2% increase in total radiation (Budyko, 1971). Over the period of time under consideration, the temperature anomaly due to atmospheric transparency variations (the temperature anomaly is equal to the difference between the values presented as curves 2 and 3) is $+0.24°$. The ratio of the second value to the first (ΔT_1) amounts to $1.2°$ per one percent.

From Fig. 5.3 it follows that the value of this ratio for shorter time intervals is less than $1.2°$ per one percent; this can be explained by the influence of the earth–atmosphere thermal inertia.

To study this feature quantitatively it is possible to use an equation given in Chapter 4:

$$-\partial T/\partial t = \lambda(T - T_r), \tag{5.10}$$

where $\partial T/\partial t$ is the rate of change with time of the mean air surface temperature T for the Northern Hemisphere; T_r the radiative temperature corresponding to the radiation income; and λ an empirical parameter depending on the earth–atmosphere thermal inertia, which for global conditions is mainly determined by the process of heat exchange in the upper layers of ocean waters.

From (5.10), provided that $T = T_1$ at $t = 0$, it follows that

$$T - T_r = (T_1 - T_r)e^{-\lambda t}. \tag{5.11}$$

Note that (5.10) and (5.11) could be used for calculating variations in both absolute temperature values and their anomalies.

The value of λ obtained in Chapter 4 from data on the annual temperature cycle in the Northern Hemisphere appeared to be 0.4 yr^{-1}. Temperature variations for longer-term periods should be characterized by smaller λ values, since in these cases the processes of heating and cooling involve deeper layers of ocean waters. This question has been studied by Oliver (1976) who found the λ coefficient to be about $0.1–0.2 \text{ yr}^{-1}$ over several decades.

It should be emphasized that the dependence of the λ coefficient on time is one of the causes of the inaccuracy of (5.11), which gives only an approximate idea of the regularities of changes in the thermal state of the earth–atmosphere system.

Taking into account the aforementioned values of λ, we can find from (5.11) the difference between the ΔT_1 parameter calculated from data on climatic changes over a 25–30-yr period and this parameter for stationary conditions. Considering $T_1 = 0$ and assuming t to be half of the time interval given above, we can find the ratio of these values to be about 0.8. It is likely that the difference of this value from unity explains why ΔT_1 obtained from data on climatic changes is smaller than ΔT_1 values determined for stationary conditions.

Keeping in mind the above dependence of air temperature on CO_2 mass growth, and using (5.11) for calculation of the temperature effects of direct radiation fluctuations, we can compute the secular trend of the mean air temperature for the Northern Hemisphere (Budyko, 1977b). For this we assume that $\Delta T_1 = 1.5°$, the relation of total radiation anomalies to direct radiation anomalies is $\frac{1}{6}$, and $\lambda = 0.1 \ yr^{-1}$.

The calculation of temperature anomalies by this method has been carried out for every year of the period under consideration, T_1 being taken from observational data for the first year. For successive years it corresponded to T obtained for the previous year. The results are depicted in Fig. 5.4 as curve 2. Comparison of this curve with observational data (curve 1) shows their great similarity.

From Fig. 5.4 it follows that the correlation coefficient between the values indicated above is 0.73. This value is sufficiently high if we take into account that the data used are approximate and the numerical model applied is schematic.

There is no reason to doubt that the actual correlation between variations in the mean Northern Hemisphere temperature and the factors taken into consideration is characterized by a larger correlation factor than that obtained here. Among the reasons for the lower correlation coefficient computed here is the lack of completely representative radiation data, as noted above, due to an insufficient number of stations for a part of the

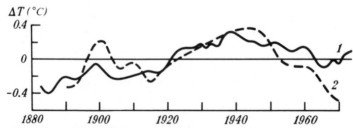

Fig. 5.4 Mean temperature anomalies in the Northern Hemisphere: curve 1, observed data; curve 2, computed results.

period in question. In addition, these data seem to reflect for recent years not only global variations in stratospheric transparency but also anthropogenic fluctuations in tropospheric transparency, which, as noted above, occur in certain regions. Since this effect is accounted for only approximately in the computation set forth here, it could lower the accuracy of the results obtained for a certain period of time.

It should be emphasized once again that the model used for calculating temperature variations is highly schematic, but it has the advantage that it incorporates a very small number of empirical parameters. It should also be mentioned that the results obtained are slightly sensitive to variations in these parameters (Oliver (1976) has drawn a similar conclusion). It is obvious that this model could be appreciably improved. This would lead to better correspondence between the curves presented in Fig. 5.4.

We can conclude that air temperature variations averaged over 5-yr periods in the Northern Hemisphere are almost completely determinated and depend on casual factors only slightly. The secular temperature trend can be computed with sufficient accuracy using simple climatic models, if major factors affecting temperature changes are taken into account. Therefore, it is possible not only to compute past climatic changes but to forecast future ones.

The second conclusion is that mean air temperature variations in the Northern Hemisphere have been considerably affected by the growth of CO_2 concentration over the last century.

Miles and Gildersleeves (1977) have also concluded that CO_2 concentration growth is important in variations in the atmospheric thermal regime over the last hundred years.

Other factors of climatic change

The influence of other anthropogenic factors on present-day global climatic conditions is rather limited. Of these factors, the production of energy in different kinds of economic activities deserves attention. This factor results in additional heating of the atmosphere and the earth's surface.

We calculated (Budyko, 1962a) the quantity of heat produced by man's economic activities. It amounts to 0.01 kcal cm^{-2} yr^{-1} for the entire earth's surface, and to 1–2 kcal cm^{-2} yr^{-1} for highly developed industrial regions with areas of tens to hundreds of thousands of square kilometers. Over large cities (tens of square kilometers) this value increases up to tens and hundreds of kilocalories per square centimeter per year. It can be calculated how this additional heat release influences the mean temperature of the earth.

It was mentioned above that a 1% change in the energy coming to the earth from the sun changes the mean temperature near its surface by 1.5°. Assuming that the heat produced by man now comprises about 0.006% of the total radiation absorbed by the earth–atmosphere system, we find that an increase in mean temperature corresponding to this quantity of heat is approximately 0.01°. This value is comparatively small. However, in some regions, it must be significantly greater because of the uneven distribution over the surface of the earth of the heat sources created by man.

It was mentioned in the above paper that, in the absence of atmospheric circulation, in highly developed industrial regions the temperature could increase by 1°, and in big cities by tens of degrees. This would probably make life impossible there. The influence of atmospheric circulation considerably weakens the corresponding temperature rise, this attenuation being greater the smaller the area over which the production of additional heat energy is concentrated.

Flohn (SMIC, 1971) has estimated the consumption of energy as an additional source of heat for the atmosphere. He established that in the central part of New York City and in Moscow the heat income produced by man exceeds by several times the amount of energy coming from the sun. In a number of smaller cities and in highly developed industrial regions with areas of tens of thousands of square kilometers the inflow of additional heat is 10 to 100% of the solar energy income.

In many countries occupying areas of hundreds of thousands of square kilometers the additional heat inflow amounts to ~ 1% of the solar radiation energy.

Flohn's data corroborate the above conclusion that not only in large cities but over large areas of industrial regions energy production by man is an important climate-forming factor.

In the first section of this chapter it was indicated that the destruction of vegetation cover and the irrigation of arid lands influence local changes in surface air temperature. It was mentioned that global climatic changes related to these measures can be opposite in character to the local changes. Thus, if irrigation causes a reduction of air temperature over fields, the mean surface air temperature will increase because of decreased global albedo due to irrigation. The reverse effect can take place as a result of the destruction of vegetation cover.

Sagan *et al.* (1978) have considered the question of variations in mean air temperature due to anthropogenic effects on vegetation cover. Different forms of man's impact on vegetation cover are considered in this paper. The authors have estimated variations in global albedo and in mean surface air temperature caused by these effects. Of greatest importance, the authors have concluded, is the anthropogenic destruction of savanna vegetation and the conversion of the savanna to desert (this occurred during several thousands of years over an area of 9×10^6 km² and was followed by an increase in the surface albedo of 0.19) as well as the felling of tropical forests (over an area of 7×10^6 km² with an increase in albedo of 0.09). Because of these processes, the global albedo of the earth–atmosphere system has increased by 0.006, resulting in a lowering of the mean air temperature by approximately 1°.

It is assumed that over the last 25 yr, as a result of the causes indicated, the albedo has increased by 0.001, producing a temperature reduction of 0.2°.

The authors of the paper believe that the temperature variations they have found explain the cooling that occurred in the little ice age as well as the temperature lowering that took place from 1930 to 1970.

It does seem that, although the estimates given here are somewhat overstated (the actual values of albedo variations seem to be lower than the values in the calculations), the effect of albedo variations on the thermal regime as indicated here has undoubtedly taken place. However, it should be kept in mind that the destruction of forests was accompanied, as indicated above, by the release of a considerable amount of CO_2, which led to increased mean air temperature. Hence, the effect on albedo variations of destroying vegetation cover was to a considerable extent compensated for.

We shall also give an estimation of the effect on the atmospheric thermal regime of variations in atmospheric CO_2 concentration caused by the destruction of vegetation cover. In the paper mentioned above, it was supposed that the total area of forests destroyed by man amounts to 15×10^6 km², a forest area of 2.5×10^6 km² having been cut during the last 25 yr. It follows from data given in Sections 5.2 and 5.3 that during this period the atmospheric CO_2 concentration grew by 7% and this increased the mean air temperature by 0.3°. If $\frac{1}{3}$ to $\frac{1}{2}$ of the total CO_2 mass produced by man is formed by the destruction of vegetation cover, the process indicated must have caused a 0.1 to 0.15° rise in air temperature over the period examined. Similar calculations carried out to assess the results of destroying forests of an area of 15×10^6 km² give a corresponding temperature rise of 0.6–0.9°.

The estimates of temperature change obtained here differ but slightly from the values of temperature lowering due to increased albedo given by Sagan and co-workers. Thus it is possible that the effect of CO_2 concentration growth has compensated, to a greater or lesser degree, for the effect of increased albedo. Unfortunately, the accuracy of the estimates considered here is insufficient for more or less reliable assessment of the total values of temperature variations caused by the destruction of vegetation cover.

One might think that the indicated temperature differences are not negligibly small and they should be taken into account in studying climatic change. At the same time, it is unlikely that this factor could be the main reason for the cooling during the little ice age and in the middle of our century.

In a study of the impact of irrigated lands on global climate (Budyko, 1971), the mean surface air temperature effect of the existing irrigation system has been estimated. The earth's surface albedo can be lowered by approximately 0.10 over irrigated zones. Considering the relationship between the albedos of the earth's surface and the earth–atmosphere system (refer to Chapter 3), we find that for little cloudiness, this decrease in the earth's surface albedo corresponds to a decrease in the earth–atmosphere system albedo by 0.07. The area of irrigated lands is approximately 2×10^6 km², that is, about 0.4% of the earth's total surface area. Irrigation thus decreases the earth's albedo by approximately 0.0003.

In Chapter 3 it was established that a 0.01 change in the earth's albedo changes the mean surface air temperature by 2.3°. Keeping this in mind, we find that irrigation increases the mean surface temperature by approximately 0.07°. This change in temperature is comparatively small.

The mean surface air temperature may have been somewhat affected by artificial reservoirs.

The mean value of the earth's surface albedo is decreased in approximately the same way by the construction of reservoirs in regions covered with vegetation as it is by irrigating deserts. Since large artificial reservoirs are built in regions with comparatively moist climate, where more or less considerable cloudiness exists, the earth–atmosphere system albedo in this case changes less than in irrigated areas with little cloudiness. In addition, since the total area of artificial reservoirs is significantly less than the area of irrigated lands, the effect of the former on the mean surface air temperature proves to be comparatively small.

The urbanization process exerts some influence on climatic conditions as a result of replacement of forests and fields by construction, asphalted roads, and so on. Sagan et al. (1978) have assessed this effect. They believed that this process had spread over an area of 10^6 km², decreasing its albedo from 0.17 to 0.15. The corresponding decrease in global albedo amounts to 2.5×10^{-5}. This increases the mean air temperature near the earth's surface by approximately 0.005°, i.e., a comparatively small value.

In addition to the processes enumerated above, the release of heat by people in the course of their vital activities may have a certain effect on air temperature. The role of this heat inflow can be felt over limited areas that are overcrowded but is insignificant over large territories.

The material in this section shows that changes in global climate during the last decades seem to depend to a certain extent on man's economic activities. This might explain the fact mentioned in Chapter 4 of the attenuation during this period of the relationship between variations in the mean global temperature and volcanic activity, which have been established from observational data for previous years.

We may conclude that further development of economic activities will lead to far larger variations in global climate.

Chapter

6

The Climate of the Future

6.1 CLIMATIC CONDITIONS IN THE NEAR FUTURE

Natural climatic change

As already mentioned in Chapter 1, for practical use, the most important data are those on climatic conditions for the next 50 yr, i.e., the late 20th and early 21st centuries.

The physical mechanism of natural climatic change throughout the last several decades has been considered in Chapter 4. According to our views, aerosol fluctuations in the lower stratosphere are generally responsible for these changes. Since aerosol fluctuations are strongly dependent on volcanic eruptions, the level of volcanicity affects climatic conditions over time periods of years and decades.

Presently there are no reliable methods for the prediction of volcanic activity. Therefore, the potential effects of this factor on the future climate can be considered only in terms of the possible extent of relevant climatic change.

As mentioned in Chapter 4, the warming trend in the early 20th century was caused to a great extent by fluctuations in volcanic activity that resulted in an increase of about 0.5° in the mean global temperature. According to Lamb (1970a), the intensity of volcanic activity fell off sharply during the 1920–1930s, but this was a relatively rare phenomenon in the regime of volcanism during the last centuries.

A quantitative analysis of Lamb's data on volcanic activity and information on secular variations in the mean global temperature enable us to draw the following conclusion. At the end of the current century the probable change in the global temperature averaged over a decade due to volcanic fluctuations will be 0.1–0.2° and the maximum change several tenths of a degree. These changes, which are not negligible, can definitely be of practical importance, mainly because precipitation in certain areas is highly sensitive to changes in the mean global temperature. However, it is clear that at the end of the 20th century and particularly in the first quarter of the 21st century the natural changes in the mean temperture will be far smaller than the anthropogenic changes.

Man-induced changes in the atmospheric composition

The previous chapter considered the idea that in the 20th century the climate is being more and more affected by an increase in CO_2 concentration in the atmosphere caused by the combustion of fossil fuels and by the reduced quantity of carbon present in living organisms and their waste products.

In recent years several studies have been made concerning the future development of energetics, which affects the global CO_2 balance (Baes *et al.*, 1976; National Academy of Sciences, 1977; Olson *et al.*, 1978).

It is quite understandable that the reliability of predictions of the progress of energetics diminishes rapidly the farther away the future they refer to. A prognosis can be realistic for 2000–2025 but it is difficult to predict the energetics prospects with reliable accuracy for later periods. Although fossil fuel reserves are sufficient for hundreds of years, the fuels within easy reach will run out and the less accessible sources will come in to use. This will result in a rise in prices for fuel and promote extensive application of nuclear and other kinds of energy that are not associated with additional CO_2 production. The available estimates of long-term energetics development are to a certain extent invalid because of difficulties in predicting technological progress in this field. Therefore, instead of prognoses for energetics development referring to the mid-21st century and later periods, it is more reasonable to consider a few scenarios and evaluate possible variations in natural conditions for each scenario. Such estimates must be taken into account in plans for energy development, in which it is desirable to envisage extensive applications of methods of energy production that are less dangerous to the environment.

It must be mentioned that for the more distant future the accuracy of calculations of the atmospheric CO_2 concentration variations is limited. This is one more reason why the estimates given by a number of authors of possible CO_2 concentrations in the second half of the 21st century and later periods are very approximate.

There are various models of the present carbon cycle that allow us to calculate future concentrations of CO_2 in the atmosphere. Let us look at the model presented by Keeling and Bakastow (1977).

Fig. 6.1 gives a general diagram of this model, with the carbon flows between the reservoirs shown as arrows. The amount of carbon in each of the three reservoirs is divided into two parts. The carbon of the land biota is represented by two components, one characterized by a slow circulation (the residence time of carbon is of the order of a few decades) and the other by a rapid circulation (the residence time is of the order of several years).

The authors have selected four different scenarios of fossil fuel usage in the next few hundred years. The CO_2 production from fossil fuel according to these scenarios is shown in Fig. 6.2. We can see that, with different scenarios, after CO_2 production has increased by 2 to 8 times the present-

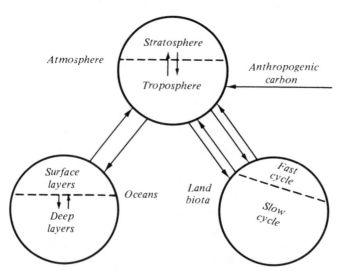

Fig. 6.1 Diagram of CO_2 circulation.

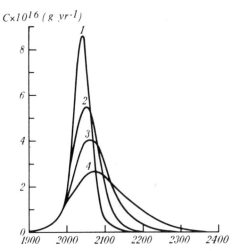

Fig. 6.2 The forecast for changing CO_2 production in the future (curves 1–4 denote different variations of progress in energetics).

day level, the rate of CO_2 release into the atmosphere decreases because of dwindling fossil fuel reserves.

To calculate future CO_2 concentrations Keeling and Bacastow used the above-mentioned model of the carbon cycle among the atmosphere, the oceans and the land biota. The present flows of carbon are based on observational data relating to the isotope content of carbon and on other empirical information.

In the calculation of the carbon balance with an anthropogenic growth of atmospheric CO_2 it is assumed that the CO_2 concentration increases only by the combustion of various kinds of fossil fuels. Photosynthetic consumption of CO_2 to increase the biota is related to the CO_2 concentration by a logarithmic function. For small variations in the CO_2 concentration this function is close to a direct proportionality between the relative increase in CO_2 consumption and its concentration variations. The coefficient of proportionality in this case is assumed to be 0.266 until the year 2000. From 2000 to 2025 it decreases by 4% a year and after that time it is assumed to be zero, which corresponds to a stable amount of biota.

The results obtained by Keeling and Bacastow are presented in Fig. 6.3. It shows the temporal variation in the CO_2 concentration as multiples of the preindustrial atmospheric concentration of CO_2. In all cases the CO_2 concentration increases by 6 to 8 times in the interval from the end of the 21st century to the end of the 23rd, depending on the rate of fuel consumption.

It is noteworthy that in the future the CO_2 concentration from the use of

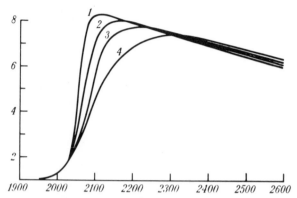

Fig. 6.3 The forecast for changing the amount of CO_2 in the future (curves 1–4 denote different variations of progress in energetics).

fossil fuel reserves will be about 0.2%. This corresponds roughly to the average level for the Phanerozoic, as shown in Chapter 2. The question arises of how long such a high concentration can be maintained in the contemporary epoch. Calculations on the basis of the available models show that the concentration will decrease but at a slow rate.

The results of the above calculation depend on an assumption concerning variations in the biota over the period in question. If we assume an unlimited growth of the biota with an increasing CO_2 concentration, the maximum CO_2 increase will be far less than we supposed it to be. Such a result is obtained, for instance, by Revelle and Munk (1977) who studied the CO_2 balance with various assumptions relating to variations in the biota on the continents.

We should like to emphasize that, if the global amount of biota decreases within the next few decades because of clearing of forests, or if it remains the same, the rate of CO_2 growth could be even greater than Keeling and Bacastow supposed.

Figure 6.3 shows that all the scenarios of energy development give similar concentrations for CO_2 in 2000. Estimates for this time obtained by other authors based on different models of the CO_2 balance differ very little. The difference increases between the results of calculations for later periods. Therefore, in evaluating climatic conditions of the future we use data on atmospheric CO_2 content only up to 2025.

Conclusions concerning CO_2 concentrations in the very near future, based on various research papers, are shown in Table 6.1.

We think that the future climate is less sensitive to other anthropogenic factors affecting the atmospheric composition than to CO_2 fluctuations,

Table 6.1
Carbon Dioxide Concentration (%)

Reference	2000	2025
Machta (1973)	0.0375–0.040	—
Stumm (1977)	0.038–0.040	0.052–0.064
Keeling and Bacastow (1977)	0.037–0.039	0.047–0.074

since there are more opportunities to restrict other factors in comparison to CO_2 production.

For example, it is hardly possible that the anthropogenic aerosols will increase considerably in the future, for this is incompatible with the preservation of favorable sanitary conditions in populated areas and with protection of the natural environment. It is well known that even now many countries take drastic measures against atmospheric pollution and efforts in this field are sure to increase in the future.

It was mentioned in the previous chapter that future increases in relatively minor atmospheric components such as Freon, nitrogen oxides, and other gases resulting from man's economic activities might also affect the climate (Ramanathan, 1976; Wang *et al.*, 1976). Evaluating their potential effects on the mean air temperature, Flohn (1978) has supported the supposition that this process will make the warming trend caused by the CO_2 rise more intense. It is possible, however, that the effect of these gas components upon the thermal regime will be less than that of CO_2.

It should be taken into account that the future climate can be affected by the heat produced from all kinds of energy consumed by man. It is evident that this effect will add to the global warming, its extent being determined by the energy balance of man's economic activities. This problem is treated below.

The thermal regime change

Considering the problem of climatic conditions in the next few decades, it can be concluded that they will depend mainly on the anthropogenic rise in atmospheric CO_2 concentration. According to the data presented in the previous chapters, doubling of the CO_2 concentration in comparison with that of the preindustrial period could occur within several decades and could increase the average temperature by about 3°. There are no other factors, either natural or anthropogenic, that could induce such great fluctuations of the thermal regime within the next 50 yr.

Figure 6.4 depicts the expected man-made variations in the mean air

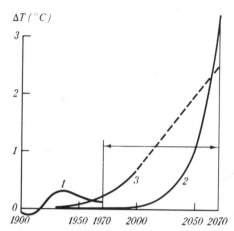

Fig. 6.4 Forecast (in 1972) of expected variations in mean air temperature at the earth's surface: curve 1, secular course of mean temperature; curve 2, temperature variations due to energy production growth; curve 3, temperature variations due to CO_2 concentration growth.

temperature near the earth's surface based on the prediction made at the beginning of the 1970s (Budyko, 1972a). In this study it was supposed that an increase in the CO_2 concentration in the atmosphere would be a basic factor in climatic change at the end of the 20th century and in the first half of the 21st, which, by the middle of the 21st century, will lead to a mean temperature rise of 2°. A total growth of heat income due to energy production could intensify the warming in the second half of the 21st century. The same prediction proposed a decrease in the meridional temperature gradient that will accompany the warming and a consequent worsening of moisture conditions in some continental areas in middle latitudes.

This prediction was introduced at the time when the climate was believed to be changing toward a cooling trend. As mentioned above, there were over 20 prognoses on climatic change in the early 1970s, each of them favoring the continuation of the cooling process. Figure 6.4 shows that, according to our prediction, the cooling was supposed to have stopped at the end of the 1960s. This supposition could not have been verified earlier than the mid-1970s, after enough observational data had been accumulated. Chapter 5 presents information that shows that at the end of the 1960s the cooling had actually stopped and possibly changed to a warming process.

The second prediction of anthropogenic climate change, which appeared a few years after the first one (Budyko and Vinnikov, 1976), was based on the data from the studies of Manabe and Wetherald (1975) and Schneider (1975).

The first of these works, which has already been mentioned, presents a calculation of the air temperature change with increasing CO_2 concentration in the atmosphere, based on the model of a general climatic theory that takes account of the general atmospheric circulation. This theory was in many respects more detailed than those that were used in the previous studies of CO_2 effects on the thermal regime of the atmosphere. For instance, it includes water exchange in the atmosphere as well as a feedback of the atmospheric thermal regime and the ice–snow boundary that determines the reflection of solar radiation in high latitudes.

Manabe and Wetherald have established in their calculations that an increase of CO_2 from 0.03 to 0.06% leads to an $\sim 3°$ rise in the mean air temperature near the earth's surface.

Schneider has considered the sources of errors in various calculations concerning the CO_2 effect on the air temperature. On the basis of joint numerical experiments with Manabe and Wetherald, Schneider has concluded that inaccurate climatic models used in the previous calculations are mostly responsible for the lower estimates of air temperature variations due to fluctuations in CO_2. At the same time he mentioned that a comparatively rough parametrization of the atmospheric longwave radiation used by Manabe and Wetherald could have led to overestimating the air temperature rise by approximately 0.5° for doubling of the CO_2 concentration.

Considering the indicated error, we assume the temperature change near the earth's surface due to doubling the CO_2 as 2.5° which corresponds to Manabe and Wetherald's results with due regard for Schneider's correction.

It was highly arguable that estimates of the CO_2 effect on the thermal regime from studies of climate theory could be used under actual atmospheric conditions, since they were obtained irrespective of the relationship among fluctuations in CO_2, air temperature, and cloudiness. It was indicated in Möller's study (1963) that air temperature variations due to CO_2 could be quite different if we consider this relationship. Many subsequent investigations are subject to the same considerations, the above-cited works by Schneider and Manabe and Wetherald as well as SMIC (1971) among them.

However, results concerning the heat income–air temperature relationship and its dependence on variations in cloudiness (see Chapter 3) show that the latter is not very strong. This enables us to utilize estimates of the CO_2 effect on temperature that are obtained with an assumed constant cloudiness.

Callendar in the above-mentioned study has established that the CO_2 effect on the air temperature is nonlinear and decreases with increasing

CO_2 concentration. We should take this effect into consideration when using the above estimates of temperature variations due to doubling the CO_2 in calculations relating to the effects on the air temperature of relatively small deviations of CO_2 from its present concentration.

According to Manabe and Wetherald's data (1967; SMIC, 1971), we can find the ratio $(\Delta T/\Delta c)':(\Delta T/\Delta c)$, where ΔT is the difference in mean surface temperatures with a Δc change in CO_2, $(\Delta T/\Delta c)$ refers to a doubling of the present CO_2 concentration, and $(\Delta T/\Delta c)'$ to an $\sim 20\%$ increase in its contemporary value. This ratio appears to be ~ 1.5.

On the basis of this estimate it has been found that within the 1970–2000 period the mean air temperature near the earth's surface will rise by $0.65°$ with a 17% increase in CO_2 concentration. Almost the same result can be obtained by applying the semiempirical theory of the atmospheric thermal regime.

On the basis of Machta's calculations concerning CO_2 changes, it has been concluded that during the same period the mean air temperature rise due to CO_2 increase should vary in accordance with the data shown in Fig. 6.5 (curve T). Curve A shows secular variations in temperature before 1975 (Budyko and Vinnikov, 1976).

According to calculations based on different climatic models with an increase in CO_2, the temperature at high latitudes varies more than that at low latitudes. For instance, the study based on the semiempirical theory of the atmospheric thermal regime (Budyko, 1974) and the works of Manabe and Wetherald show that the temperature change within the 60°–90° latitudinal belt is 2.5 times the mean global value.

More detailed information concerning the future thermal regime has

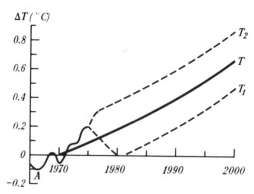

Fig. 6.5 Forecast of mean air temperature variations (1976): T, probable air temperature change in the Northern Hemisphere until 2000 relative to the temperature averaged over the 95-year period 1881–1975; A, temperature change from observed data.

been obtained using not only calculations based on climate theory but empirical data on regularities of climatic changes at present and paleoclimatic evidence (Budyko *et al.*, 1978b). It was assumed in this study that CO_2 concentration would be 0.037% in 2000 and 0.065% in 2025. We think that these values correspond to the highest level of possible variations in CO_2 concentration. Therefore, the climatic change as estimated for such conditions should be considered the greatest possible for these periods of time.

With the help of the above method it was found that in 2000 such an increase in CO_2 would lead to a 0.6° rise in the mean global temperature. In 2025 the CO_2 concentration in the atmosphere will be approximately twice that of the 1970s. This will correspond to the chemical composition of the atmosphere during the Pliocene when the mean temperature in the Northern Hemisphere exceeded the present one by 4.8°. If we assume that polar glaciations and aridity of the continental surface increase the earth's albedo and decrease the present mean air temperature near the earth's surface by about 2.0° compared to the Pliocene, we find that doubling the CO_2 will result in a 2.5–3° temperature increase. This agrees well with similar estimates based on climatic theories.

The above calculations of a forthcoming warming did not take into account the effect of oceanic thermal inertia on air temperature variations. Using the model presented in Chapters 4 and 5, it can be found that the indicated effect leads to a slowing down of the warming process on the order of 10 yr, i.e., by a value of the order of the accuracy of the calculations. It is quite likely that this value will be compensated for by the effect on the thermal regime of additional factors related to fossil fuel combustion, for instance, trace gases absorbing longwave radiation.

If by 2000 a warming trend leads only to a relatively small contraction of the polar sea ice area, its further intensification might cause permanent sea ice to disappear before the end of the first quarter of the 21st century. This conclusion is based both on the calculated variations of the heat balance of ice during a warming trend (Budyko, 1971) and on the analysis of the empirical relationship between the present boundary of the sea ice and the temperature of the lower air layer.

The area covered by sea ice and the average thickness of the ice depend mainly on climatic conditions. The previous work (Budyko, 1971) introduced a numerical model based on the equations of heat and water balance of sea ice that enables one to calculate the present boundaries and thickness of sea ice. The results obtained agree reasonably well with observational data (Budyko, 1969a, 1971).

According to our calculations based on this model, the ice regime depends on the solar radiation income, the air temperature during warm periods of the year (ice melting season), and the air temperature of the cold season when ice thickness increases mainly because of water freezing at the lower surface of the ice. Results have shown that the ice thickness varies greatly in the warm seasons, even with relatively small air temperature fluctuations, and it is far less affected by temperature variations during the cold season.

Utilizing these results, we can estimate when central Arctic many-year ice will turn into one-year ice because of the progression of global warming.

According to the empirical data on present climatic change, the summer temperature rise in the Arctic sea-ice zone is on the average 100% of the mean temperature increase in the Northern Hemisphere. In winter the temperature change there is 4 to 5 times greater than that. Therefore, a 2° increase in the mean temperature for the hemisphere corresponds to a temperature rise of 2.0° in summer and 8–10° in winter in the zone of sea ice.

Calculations based on the above theory show that under such conditions the average decrease in ice thickness in summer exceeds 3 m. This is enough to convert many-year ice into one-year ice. After that, additional warming must occur due to a decrease in the earth's surface albedo in high latitudes. In a number of studies it has been concluded that in the central Arctic the ice-free regime can exist even without anthropogenic heating of the atmosphere (Budyko, 1971). Therefore it is quite possible that, with this heating at work and with many-year ice melted, one-year ice will either disappear or persist only in the limited off shore regions of the Arctic Ocean with the most severe winters.

In addition to the semiempirical method we have used to evaluate conditions for sea-ice melting when a warming process is going on, there is a simple empirical solution of this problem.

We should remember that many-year Arctic sea ice exists at an air temperature in the warmest month of from − 1 to + 2°C, the major part of the ice cover being in the zone where this temperature is from 0 to + 2°C. Studies relating to modern climatic change have established that during a warming period the air temperature rise close to the sea-ice margins is somewhat higher than the mean temperature increase over the entire ice-covered zone. According to the empirical data, the summer temperature rise there is more than 100% of the mean air temperature increase in the Northern Hemisphere.

In other words, if the temperature averaged over the hemisphere rises by 2°, the many-year ice that covers a zone with a warmest-month temperature of 0 to + 2°C would melt, first at the edges where the temperature rise would be the greatest, and then in the central regions. Evidently, as the ice melting proceeds, the air temperature will rise in the regions where it used to be somewhat below 0°C during the warmest month. This would be the result of additional warming due to a decrease in the albedo of the earth's surface, which would lead to the complete disappearance of the ice cover. Taking into account that the mean air temperature has already increased by a few tenths of a degree as compared to the end of the last century, and allowing for possible error in the calculations, we believe that, if the mean global temperature rises to 2–3° higher than in preindustrial time, it will be enough for the many-year ice in the Arctic to disappear. Such an increase in temperature can be expected even before 2025. It is obvious that the time for ice melting depends on the acceleration or slowing down of the warming process due to atmospheric transparency fluctuations. As mentioned above, once the sea ice has melted, an additional temperature rise will occur, which would be especially significant in high latitudes.

Climatic conditions to be expected in the USSR in 2000 as a result of anthropogenic warming may be predicted by an empirical model of present-day climatic change developed from observational data covering the period of the existence of the world meteorological network (Vinnikov and Groisman, 1979).

To describe global climatic change, an annual mean air temperature for the Northern Hemisphere was used in this paper. A statistical method was utilized to determine relationships among variations in the local parameters of the thermal regime in different seasons of the year and fluctuations in the mean air temperature. Table 6.2 presents estimates of the possible variations in zonal air temperature for different seasons. To obtain these values it was assumed that the mean annual temperature would rise by 0.5° in the Northern Hemisphere. The indicated estimates may be considered a prediction of the air temperature variations by 2000 relative

Table 6.2

Variations in the Surface Air Temperature (°C)
at Different Latitudes of the Northern Hemisphere
by 2000

Season	North latitude				
	20–40°	45–60°	65°	70°	75–85°
Winter	0.4	0.6	1.4	2.0	2.4
Spring	0.3	0.5	0.6	0.6	0.6
Summer	0.2	0.4	0.8	0.8	0.6
Fall	0.3	0.6	1.0	1.1	1.1

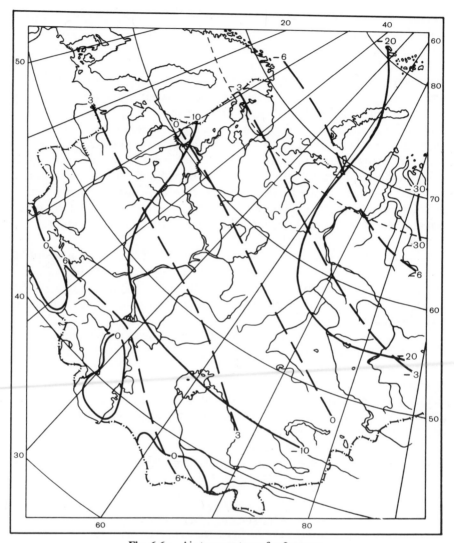

Fig. 6.6a Air temperatures for January.

to the mean air temperature in the 1970s. It is seen that, if the mean air temperature rises by 0.5°, the most considerable change in the thermal regime will occur in winter in the Arctic, where the mean air temperature will rise by 2–2.5°. The air temperature increase at all latitudes in summer and at low latitudes in all seasons will be several tenths of a degree.

One should bear in mind that such temperature variations might take

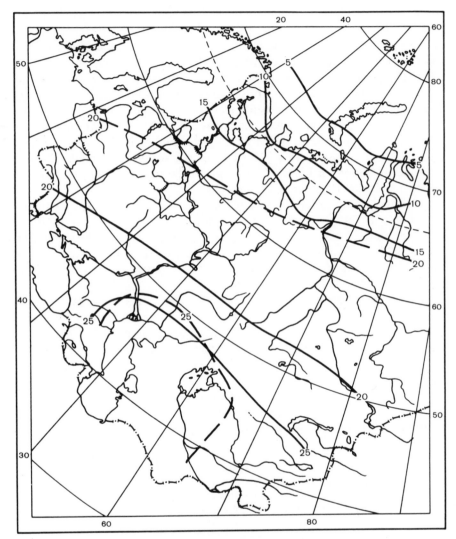

Fig. 6.6b   Air temperatures for July.

place if the atmospheric transparency approaches its normal state. A higher transparency increases the warming trend, a lower one, which could be a result of more intensive volcanic activity, slows down the warming process.

As already mentioned, in the 2020s the thermal regime will approach that of the Pliocene if the amount of CO_2 doubles. For the latter period palaeoclimatic maps are available for Europe and Asia, drawn by Sinitsyn (1965, 1967) and based on the analysis of lithogenetic, geobotanic, and zoogeographic data. Schematic climatic maps for the USSR territory (without eastern Siberia) are presented in Fig. 6.6, based on Sinitsyn's data. The dashed lines show isotherms for January (Fig. 6.6a) and July (Fig. 6.6b) characterizing the possible climatic conditions in the 2020s. The solid lines represent the isotherms for the same months for the contemporary epoch (Atlas, 1964). It is seen from these maps that by the 2020s the air temperature in the Arctic area would rise by 12–15° in July and 15–20° in January. In middle latitudes it would increase by 2–5° in July and by 10–15° in January. The zero isotherm for January would shift northward by 10–15 degrees of latitude.

It follows from the data on the warmest and coolest months' temperature that in 45–50 yr the thermal conditions of the northwest of the European part of the USSR would be similar to present-day climatic conditions in central France, those of the north of West Siberia would correspond to the southern part of Poland, and those of the central regions of West Siberia to the Middle Danube lowland.

The temperature rise will increase the duration of the period with above-zero air temperature and of the growing season. Table 6.3 presents data on variations in 2000 and 2025 in the total of the mean diurnal air temperature over a period when the temperature is higher than 10°C, which is one of the main agroclimatic characteristics (these sums are equal to the product of the mean air temperature on the Celsius scale over a period when the mean diurnal temperature is over 10°C and of the length of the period). Throughout the area under consideration the sums of the air temperature in 2000 would rise by 100–300°.

Table 6.3
Air Temperature Sums (°C) in 2000 and 2025 (for Periods with
Temperatures Higher Than 10°C)

North latitude (degrees)	East longitude (degrees)								
	30			50			70		
	At present	2000	2025	At present	2000	2025	At present	2000	2025
66.5	1000	1300	2400	700	900	2500	650	950	2500
60	1650	2000	2900	1550	1700	3000	1500	1650	3000
55	2100	2300	3400	2250	2400	3500	2000	2150	3400
50	2600	2750	3750	3100	3300	4500	2400	2550	4000
47	3200	3300	4500	3400	3550	4900	2900	3050	4500

In 2025 the sums of the air temperature would increase by 1400–1800° at the polar circle latitude, by 1200–1600° in the latitudional zone from 50 to 60°N, and by 1300–1600° in the south of the territory in question.

It follows from Table 6.3 that the thermal zones that are described by the growing season conditions would shift northward in 2000 by 1–3 degrees, and in 2025 by 10–15 degrees of latitude in comparison with present-day conditions (Budyko *et al.*, 1978a).

These data on the future thermal regime can be compared to Kellogg's estimates (1977, 1978) concerning variations in the mean global temperature due to increasing CO_2 concentration in the atmosphere relative to preindustrial times. His estimates for the current period (+ 0.5°) and for 2000 (+ 1.2°) coincide with our results. For 2025 Kellogg has estimated an increase in temperature of 2.6° compared to preindustrial times (or 2.1° relative to the modern epoch), which is somewhat lower than our result.

Pearman (1977) presents a diagram of the future mean air temperature change near the earth's surface resulting from an increase in CO_2 concentration. The diagram shows the most probable temperature change at the end of the 20th century and the beginning of the 21st, the results practically coinciding with our estimates.

The available estimates of the thermal regime within the next 50 yr have confirmed the conclusion made somewhat earlier (Budyko, 1972a) that, within this period, direct heating of the atmosphere as various kinds of energy are converted into heat would not have a significant impact on global climate. For later periods of time, opinions on this problem differ. The author's works (Budyko, 1962a, 1972a) suggest that, with a further increase in energy production, anthropogenic heating of the atmosphere will become significant in the second half of the next century.

The authors of "Energy and Climate" (National Academy of Science, 1977) think that future centuries will not see more than a 0.5° rise in the mean global air temperature due to the above factor. Since the methods applied for the estimation of the energy impact on climate in the above studies are practically identical, the difference of opinion is entirely explained by variation in the concepts concerning the rates of energy production in the distant future. This problem is difficult to solve because the investigators make different and arbitrary assumptions. However, all of them believe that the local climate in the future will be more affected by anthropogenic heating of the atmosphere.

Variations in the water regime

To estimate variations in precipitation by 2000, Vinnikov and Groisman have applied an empirical model of climatic change. This model deter-

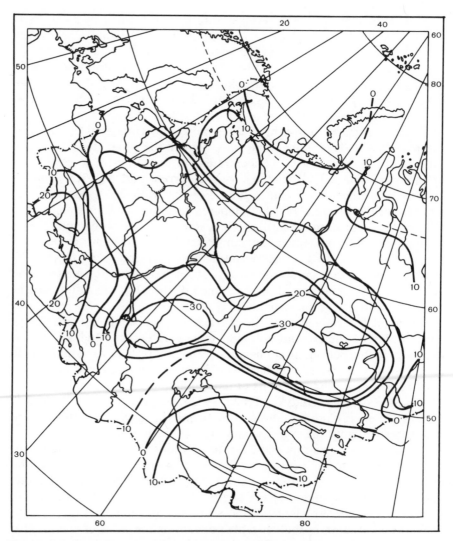

Fig. 6.7 Relative changes in total precipitation as a percentage of the normal precipitation for winter with an increase in mean hemispheric temperature of 0.5°.

mines the relationship between precipitation totals over the territory of the USSR and global warming or cooling trends. Figure 6.7 presents relative variations that will occur in the total winter precipitation if the mean global temperature rises by 0.5° (Vinnikov and Groisman, 1979).

To obtain these estimates measurements taken over the 1891–1973 period at more than 500 stations have been used. The necessary adjustments have been made according to modifications in precipitation measurement techniques to obtain homogeneous sets of observational data. The map shows that a 0.5° global temperature rise will lead to a decrease in winter precipitation over most of the steppe and forest-steppe zone of the USSR. The decrease in precipitation during this season will in some cases reach 30% of the normal value. Although the mean annual total precipitation varies less than the total precipitation for the winter season, its change reaches 10–15% in a number of regions.

The reason for such a change in moisture conditions lies in less intensive atmospheric circulation and water vapor transportation from the oceans to the continents when the meridional temperature gradient decreases, which is characteristic of a warming period (Drozdov and Grigorieva, 1963, 1971; Budyko and Vinnikov, 1973).

When, in the 2020s, a rising CO_2 concentration causes many-year polar ice to disappear, rainfall will increase due to an increase in temperature and in evaporation from the polar sea surface. Similar conditions existed in the Pliocene, as can be seen from the maps of precipitation and temperature distribution drawn up by Sinitzyn. Analysis of water exchange in the atmosphere in the late Tertiary shows a noticeable reduction in moisture flow from the oceans to the continents in comparison with the contemporary epoch. To some extent this compensates for an increase in precipitation under the influence of a temperature rise. As a result, the total precipitation increases in outlying continental areas but does not materially change in the central parts of the continents.

One can see from the map in Fig. 6.8, drawn on the basis of Sinitzyn's data, that in 2025 in the present-day tundra zone rainfall would exceed the present-day amount of precipitation by 500–600 mm a year. Southward to the Riga–Tselinograd line there will be an annual increase in rainfall of approximately 200–400 mm. To the south of this boundary the total amount of precipitation will change slightly.

Considering the above-stated ideas concerning precipitation variations and their causes, one might think that the present type of annual rainfall variations in the greater portion of the territory in question, with a summer maximum and winter minimum of rain, will be maintained in the future as well.

During a period preceding the disappearance of sea ice, a decrease in the meridional temperature gradient will lead to a reduction of atmospheric circulation stability and severe droughts will occur more often on the continents in middle latitudes. We had similar variations in the precipitation regime during the warming trends of the 1930s and the first half of the 1970s.

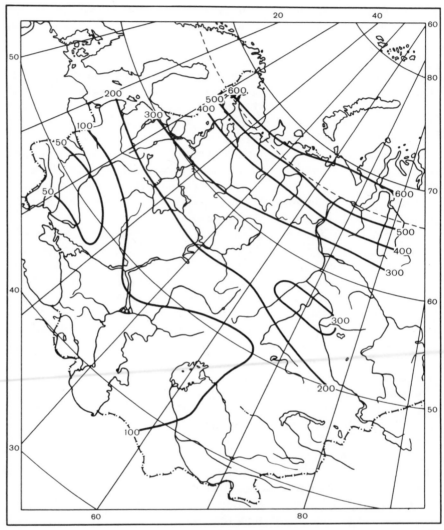

Fig. 6.8 Differences in annual precipitation totals (in millimeters) in the 2020s and at the present epoch.

As warming intensifies in all latitudes, the amount of potential evaporation from the land surface will grow under the direct influence of a temperature rise as well as an increase in the radiation balance of the earth's surface. The latter results from the CO_2 concentration rise and a decrease in the earth's surface albedo when snow cover disappears.

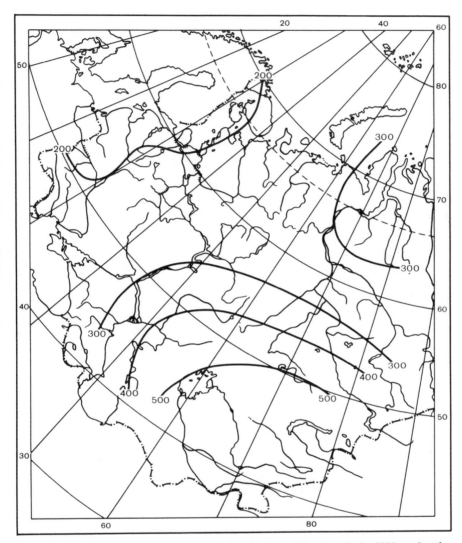

Fig. 6.9 Differences in annual potential evaporation (in millimeters) in the 2020s and at the present epoch.

The value of the radiation balance and the amount of potential evaporation expected in 2025 have been computed for 40 points distributed evenly over the territory under consideration.

When determining the radiation balance for doubling of the CO_2 amount in the atmosphere, it was assumed that the effective longwave radiation was 6% lower, according to the results obtained by Manabe and Wetherald (1975). The values of the albedo and the absorbed radiation were found, taking account of the air temperature rise, which leads to an increase in the duration of the growing season and the snow-free period. Under the influence of the indicated factors, the annual total radiation balance is 10–12% greater in the southern and central areas and it increases by 30–50% and more in the north, in West Siberia, and in North Kazakhstan.

The potential evaporation in 2025 was computed on the basis of a heat balance method (Budyko, 1971) taking account of the temperature rise and radiation balance assumed for 2025. The air humidity was determined empirically on the basis of quantities valid for similar landscapes with corresponding values of the air temperature and radiation balance.

Figure 6.9 depicts a map of the differences between potential evaporation expected in 2025 and that of the contemporary period. The data presented in the map show that in 2025 potential evaporation should increase by 200 mm yr^{-1} in the west-European part of the USSR, by 300 mm yr^{-1} in the central and northeastern regions, and by 400–500 mm yr^{-1} in the southeast of the territory under consideration.

How river runoff will vary under the changed climatic conditions was found on the basis of precipitation and potential evaporation data for 2025. Figure 6.10 presents differences (in mm yr^{-1}) between river runoff in 2025 and the present. It is seen that the runoff rises by a factor of 2 to 3 north of 58–60°N. In the central and southern regions the changes in the runoff are comparatively small and the runoff decreases somewhat in the Dnieper and the Don basins (Budyko *et al.*, 1978a).

Few other data are available on water regime changes resulting from man-induced warming. In a study estimating the future water balance Kellogg (1977, 1978) presents a map referring to the period known as the Climatic Optimum (roughly 4000 to 8000 yr ago) in which regions wetter and drier relative to the present are shown. Kellogg suggests that similar changes might occur in the patterns of precipitation if there is a warming trend caused by an increase in atmospheric CO_2.

For some regions the data on the map agree with our results; for others no agreement can be observed. Discrepancies might result from inaccurate data on past moisture conditions and differences in the climate genesis of the late Tertiary and the Climatic Optimum during the Holocene.

The next subsection presents conclusions about variations in the moisture regime due to anthropogenic warming, which were drawn from calculations on the basis of a general circulation model (Manabe and Wetherald, 1980).

It is evident that, since the future patterns of precipitation and runoff

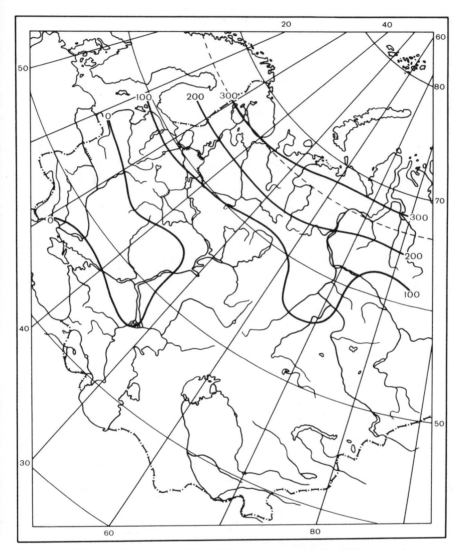

Fig. 6.10 Differences in river runoff (in millimeters per year) in the 2020s and at the present epoch.

are very important for economic activities, it is necessary to carry out detailed studies of forthcoming changes in the constituents of the water balance on the basis of numerical climate models and the appropriate paleoclimatic data.

Reliability of future climate data

It is of vital importance that the available information on the climatic conditions of the near future be reliable. This can be seen just from the list of applied problems that cannot be solved without data on future climate.

Considering this subject, we can divide it into two parts, the first of which deals with calculations of the CO_2 content of the atmosphere and the second—with the estimation of CO_2 impact on climate.

As mentioned above, the next decades will not likely witness radical changes in the nature of modern energetics, which is now widely based on fossil fuels. At the same time, a rapidly growing world population calls for higher rates of food production, which would necessarily result in the expansion of arable lands by forest clearing, among other ways. This inevitably leads to a decrease in the carbon contained in forests.

The available models of the carbon balance, based on various assumptions concerning rates of fossil fuel consumption and carbon exchange in the biota, yield quite similar results for the prediction of CO_2 concentrations in the atmosphere in the next few decades. According to all the realistic models, the atmospheric CO_2 concentration will double in a few decades. The dates for this doubling obtained by the available calculations are different, but this is of no importance since the major conclusion holds that CO_2 will double in less than 100 yr and possibly in no more than 50 yr.

It is possible to change this time span somewhat by planning long-term energy production on a global scale. In this planning, data on anthropogenic climatic change should be taken into account. To acquire these data is our main task.

It is evident that these data will be of a relative nature and refer to the realization of definite scenarios of economic progress. However, in the near future the realization of various scenarios of man-made CO_2 production is hardly possible.

As to the question of the impact of the growing CO_2 concentration on climatic conditions, there is an opinion that it can be solved only with an exact climatic theory. Since at present such a theory does not exist, and one will not be introduced soon, some scientists think that no reliable information on the future climate can be obtained for the time being. For example, this standpoint was expressed in the report of a conference on the theory of climate held in 1974 (International Study Conference, 1975).

We can hardly share such views. As is known, approximate physical theories are widely used in many branches of engineering and other applied sciences since precise theories of corresponding processes are either absent or unnecessary. To make our point clearer, we give an example from hydromechanics that is comparatively closer to the area of the atmospheric sciences. Despite outstanding achievements in the development of the theory of streamline properties of liquids and gases, approximate theories yielding quite reliable results are widely applied in designing ships and aircraft. For substantiating such theories, determining input parameters, and verifying the results obtained experimental studies are extensively used.

In our opinion it is quite natural to use a similar approach for climatic theories because a great deal of empirical data is available in this field and the data-stock is growing rapidly as experimental techniques for studying present and past climates improve.

The study of the earth's surface heat balance carried out by the author and his co-workers since the 1940s can be an example of a combined approach solving some problems of physical climatology. Until that time, neither exact values of the heat balance components nor even their sign and order of magnitude were generally known. Since the accuracy of all methods for theoretical and experimental determination of the heat balance components was limited, in our first works we suggested several independent methods for determining the heat balance components and the results were thought reliable if they agreed reasonably well (Budyko, 1946, 1947). That method proved rather effective for the study of the earth's surface heat balance. Although numerous new maps of the heat balance components were constructed in the subsequent studies (Budyko, 1955, 1963; Budyko *et al.*, 1978a), they usually differed from our original data by no more than 10–20%. So the results of the first heat balance studies seem to be reasonably reliable.

We would like to emphasize that later maps of the earth's surface heat balance components based on detailed climatic theory (Holloway and Manabe, 1971) proved to be similar to those based on the earlier semiempirical methods.

Whereas the heat balance study was of limited practical importance and more or less rapid collection of data on the heat balance constituents did not greatly affect economic activities, delay in obtaining reliable information concerning the future climate, which may be subjected to rapid changes, could have drastic economic consequences. Therefore, we should apply the methods of the modern science of climate to their fullest extent to learn about the possibilities of climatic change.

The basic problem in this field is the determination of the sensitivity of modern climate to fluctuations in the heat income. The data presented in Chapter 3 show that remarkable success has been achieved in the study of this problem. For instance, the most detailed investigations based on different methods have yielded similar results for the difference in the mean air temperature near the earth's surface for a 1% change in the solar constant. For the CO_2 impact on the thermal regime we have again quite reasonable agreement between the estimates obtained by different methods. At present most scientists believe that a doubling of the CO_2 concentration raises the temperature by about 3°. The possible error of that estimate has been argued more than once.

Since similar estimates of this parameter can also be obtained from different empirical data, we think that the probable error does not exceed 20 to 30%. This accuracy is sufficient for practical purposes.

In addition to the mean air temperature variations near the earth's surface we can determine quite accurately the mean latitudinal temperature change for a doubling of the CO_2 concentration. Different theoretical and empirical methods provide results for this change that agree closely. In order to know how the moisture regime will change with the growth of the atmospheric CO_2 concentration, it is essential to compare the above data based on paleoclimatic evidence with recent results by Manabe and Wetherald (1980) who estimated the climatic change for a twofold and fourfold increase in CO_2 using the general circulation model. An important feature of their work is that it deals with climatic conditions over land and ocean separately, although the topographies of land and ocean were highly schematic.

The estimate of the sensitivity of the mean hemispheric air temperature to variations in CO_2 practically coincides with the previous conclusion of the same authors (Manabe and Wetherald, 1975). The mean latitudinal distribution of air temperature change near the earth's surface due to doubling of the CO_2 concentration is also quite similar to their earlier estimate. Manabe and Wetherald confirmed once more that cloudiness variations due to CO_2 increase have little effect on the thermal regime.

They have also established that doubling and quadrupling of the CO_2 concentration are responsible for a noticeable increase in continental precipitation in the latitudinal belt above 50°N. In the same latitudinal belt, runoff and soil moisture are also increased. The precipitation and runoff, as well as the soil moisture, decrease over a zone centered around 45°N but in the lower latitudes they again increase. Although these conclusions hold true under average latitudinal conditions on a continent whose outlines are highly idealized, they agree reasonably well with the empirical results obtained for the USSR territory as presented in Figs. 6.8 and 6.10. These figures show that in the western part of the Soviet Union doubling of the CO_2 in the atmosphere is responsible for enhancing precipitation and runoff mainly north of 50°N.

Since the data of the maps for more southern latitudes cover a very limited area it is difficult to verify the conclusion that moisture conditions become worse in the belt around 45°N. But the area around 50°N where the runoff rate is reduced is clearly marked on the map. From Figs. 6.8 and 6.9 it can be seen that soil moisture in the southern regions of the territory in question is reduced since the increased potential evaporation considerably exceeds the enhanced precipitation. This conclusion agrees with that of Manabe and Wetherald.

In view of the reasonable agreement among qualitative results related to the change of moisture conditions in response to CO_2 increase in the atmosphere, we think that these conclusions are true, since they are based on entirely different methods.

In order to verify these estimates quantitatively we need calculated data based on climatic theory for a realistic topography, which can probably be obtained in the near future.

Concerning the effect of climatic warming on polar ice, we should emphasize the above-mentioned possibility of concluding that permanent polar sea ice might disappear in response to a doubling of the CO_2 content. This conclusion can be reached from data on the mean latitudinal temperature change by three independent methods.

We can judge the efficiency of the semiempirical climatic model used for the prediction of the future thermal regime by its successful application to the quantitative estimation of the most significant natural climatic changes that occurred in the last hundred million years, including the Cenozoic cooling, the Quaternary Ice Ages, and contemporary fluctuations in the thermal regime.

The reliability of conclusions based on the calculations of future climatic conditions according to the applied climatic model is further proved by the agreement between the results concerning past climatic change and the empirical data.

Finally, we think it is possible to recognize that an anthropogenic warming is very likely to develop within the next few decades, a warming whose scale could be compared to climatic changes occurring millions of years ago. At present only the general features of the forthcoming change are known and the accuracy of the available information concerning various elements of the future climatic conditions is uneven.

The question of the sensible practical use of information on the future climate is quite complicated.

Our paper containing the first prediction of the forthcoming climatic change (Budyko, 1972a) stated that it was still early enough to use it for planning economic activities but the results obtained required reliable information on future climatic conditions, which should be on hand before the end of a decade. By the time the book appeared, this period had expired and the urgency for clarifying the prospects for climatic change had become evident. At the same time it is clear that for successful long-term economic planning we must have precise and detailed information on the future climate, which is unavailable at present.

We are of the opinion that, faced with this situation, we must make practical use of the available information on future climatic change. Otherwise our long-term planning will be based on the concept that present climatic conditions will not vary in the future, a concept that is not shared now by most of the world's leading climatologists. We can easily understand that, since drastic climatic change should occur in the future, the above view can lead to unfavorable consequences for economic planning.

The only alternative, in our opinion, is the development of an optimum strategy for use of the available data on the future climate in planning, allowing for their limited accuracy and detail.

In some cases this strategy can be based on the information corresponding to several of the most probable versions of future climatic change. For more important tasks the choice of optimum solutions for planning must rest on specific numerical models of climatic impact on economic activity that enable us to take into account the accuracy of the climatic data used.

Considering the vital importance of recommedations relating to future climatic conditions, we think it is expedient to charge special national and international organizations with the carrying out of these recommendations.

6.2 THE NEAR FUTURE OF THE BIOSPHERE

Variations in geographical zonality

It follows from the first section of this chapter that the impact of human economic activities on climate will result in the near future in a major warming trend embracing the entire globe.

It is important to recognize that this change in climate will correspond to the restoration of climatic conditions typical of the remote past, namely, of the Tertiary period. To show this situation, Fig. 6.11a depicts variations in CO_2 concentration in the Cenozoic era, before the modern epoch. Part (b) of the figure shows the CO_2 concentration since 1900 on a larger time scale. The upper and lower curves correspond to the results of Keeling and Bacastow based on the extreme alternatives of the future development of energy production (Keeling and Bacastow, 1977).

Comparing both sides of the figure, we can easily see that after a relatively slow rise in the 20th century, the CO_2 concentration goes up sharply

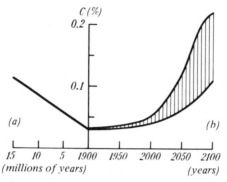

Fig. 6.11 Carbon dioxide concentration change (a) during the Tertiary and (b) at the present epoch.

in the 21st century, reaching the level of the Tertiary. This rapid restoration of the ancient atmosphere is accounted for by the fact that the resources of coal, oil, and natural gas that accumulated over millions of years are now being burned within a decade in the course of man's economic activities.

Figure 6.12 presents calculated variations in the mean air temperature near the earth's surface (a) for the Neogene and (b) for the 20th and 21st centuries. The values of ΔT correspond to differences between temperature means for different years and that at the beginning of the 20th century. The Neogene data are based on both paleoclimatic information and climatic model calculations and the results obtained proved to be in good agreement. Two methods are also used to calculate ΔT for the late 20th and 21st centuries. Model calculations and those based on the empirical relationship between ΔT and CO_2 concentration yielded similar results. It

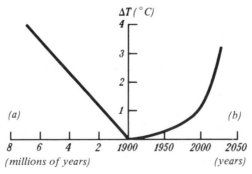

Fig. 6.12 Air temperature change at the earth's surface (a) during the Neogene and (b) at the present epoch.

can be seen from the figure that at the end of the first quarter of the 21st century the mean air temperature near the earth's surface will reach the level of the Pliocene and will continue to rise.

Model calculations and paleoclimatic data show that there are great differences in the mean temperature increases due to the growth of CO_2 concentration at various latitudes. At high latitudes the temperature rise is considerably higher than in low latitudinal belts. Therefore, with an anthropogenic climate change, the natural conditions will be modified more drastically in high latitudes and to a lesser extent in middle latitudes.

One of the most important consequences of the anthropogenic climate change will be a shift in the geographical zones due to climatic factors.

The relationship between the geographical zones of the USSR and the climatic conditions has been studied by Grigorjev and Budyko (1959), who introduced a classification of the climates of the Soviet Union worked out on the basis of the above relationship.

In the development of this classification, data on the sums of temperatures at the earth's surface, which are connected with the radiation balance, were used to define thermal conditions. In the calculation of the dryness index (equal to potential evaporation/total annual precipitation), which describes the moisture conditions, potential evaporation was determined by the heat balance method (Budyko, 1971).

A comparison has shown that the boundaries of geobotanic zones agree well with the distribution of climatic characteristics used in the classification. The region where the mean air temperature throughout the year remains below 10°C appears to correspond to that of the Arctic deserts. The isoline for the sum of surface air temperatures equal to 1000°C approaches the southern border of the tundra zone, the forest-tundra included. This same natural border proved to be close to the isoline for a dryness index equal to 0.45. The region where the dryness index is lower than 0.45 but the temperature sums are greater than 1000°C corresponds to the position of high-mountain alpine meadows. The isolines for a dryness index equal to 1 closely approach the border between forest and steppe zones. Within the forest zone the area with temperature sums under 2200°C is occupied by coniferous forest (taiga); from 2200°C to 4400°C by mixed and broadleaf middle latitude forest; over 4400°C by forests with some subtropical vegetation.

A dryness index of 1 to 3 corresponds to vegetation zones of insufficient moisture conditions, mainly steppe and forest-steppe. Under such conditions, at temperature sums above 4400°C, the steppe of the middle latitudes turns into a steppe where species of subtropical vegetation are present.

A dryness index greater than 3 corresponds to desert conditions where,

Table 6.4
Moisture Conditions

Moisture conditions	Index of dryness (potential evaporation to precipitation ratio)	Geographic conditions
1. Very moist	<0.45	Arctic desert, tundra, forest-tundra, alpine meadows
2. Moist	0.45–1.00	Forest
3. Insufficiently moist	1.00–3.00	Forest-steppe, steppe, xerophytic subtropical vegetation
4. Dry	>3.0	Desert

at relatively low temperature sums (<2200°C), mountain deserts are observed. In the zone of northern lowland deserts, at lower temperature sums (up to ~4400°C), conditions of Artemisia and saltbush deserts prevail; at greater sums, those of saxaul, shrub and ephemeral deserts.

In designing a system of climatic regionalization we have used gradations of the basic climatic indices based on the above regularities (see Tables 6.4 and 6.5).

A general schematic classification of the climates of the USSR is given in Table 6.6. It is seen from the table that twelve types of basic climatic regions corresponding to various geographical zones are characteristic of the Soviet Union.

The spatial locations of climatic zones and regions are shown as a map of USSR climatic regionalization in Budyko (1971). This map shows that

Table 6.5
Thermal Conditions of the Warm Period

Thermal conditions	Sums of surface temperatures for the period with the air temperature exceeding 10°C	Geographic conditions
1. Very cold	Air temperature <10°C throughout the year	Arctic desert
2. Cold	0–1000°C	Tundra and forest-tundra
3. Moderately warm	1000–2200°C	Coniferous forest, alpine meadows, mountain steppe, and Siberian steppe
4. Warm	2200–4400°C	Mixed and broad-leaf forest, forest-steppe, steppe, northern desert
5. Very warm	>4400°C	Subtropical vegetation, desert

Table 6.6
Climatic Factors of Geographical Zonality

Thermal conditions of warm period (sum of surface temperatures)	Moisture conditions (index of dryness)						
	1. Very moist, less than 0.45		2. Moist, 0.45–1.00	3. Insufficiently moist, 1.00–3.00		4. Dry, over 3.00	
1. Very cold, air temperature less than 10° all year round	1.1 Arctic desert						
2. Cold, 0–1000°	1.2 Tundra and forest-tundra						
3. Moderately warm, 1000–2200°	1.3 Alpine meadows	2.3 Coniferous forest		3.3 Mountain steppe and Siberian steppe		4.3 Mountain desert	
4. Warm, 2200–4400°		2.4 Mixed and broad-leaf forest		3.4 Steppe and forest-steppe		4.4 Northern desert	
5. Very warm, over 4400°		2.5 Subtropical forest		3.5 Xerophytic subtropical vegetation		4.5 Desert	

the zones of excessively moist climate with a very cold summer (1.1), corresponding to the Arctic deserts, are situated on islands in the Arctic Ocean and in the northern mountainous part of the Taimyr Peninsula. Glaciers occupy a considerable portion of this zone. Almost all year round the glacier-free area is covered by snow and its soil mantle is undeveloped (Arctic soils). Very scanty vegetation such as lichen and moss does not cover the ice-free area completely but occupies generally less than half of the total surface. Evaporation there is rather small due to low temperatures and high humidity and in some periods is replaced by condensation. Precipitation is mainly disposed of as runoff.

The zone of excessively moist climate with a cold summer (1.2) almost coincides with the tundra and forest-tundra covering the northern and northeastern coast of the USSR and some interior continental regions in the northeast (mountain tundra). The tundra soils typical of this zone usually contain little organic matter and are characterized by low activity of biological processes. The tundra vegetation which grows within a relatively brief period when snow cover is absent, consists of moss and lichen as well as shrubs and dwarf trees in the southern part of the zone. We do not know exactly the amount of annual evaporation in the tundra zone, but evidently this process consumes a minor fraction of the precipitation, probably on the average no more than 30%. Therefore, the runoff coefficient is comparatively high there.

The third climatic zone of excessive moisture conditions, with a moderately warm summer (1.3), corresponds to the regions of alpine meadows. Such conditions are observed in some mountain regions in the Soviet Union. This area exhibits a great variety of vegetation species growing in subalpine and alpine meadows on mountain-meadow soils. The precipitation ensures an abundant runoff and a high runoff coefficient.

It is typical of the landscapes of climatic zones with surplus moisture that forest vegetation is absent. We should mention that forests are absent not only because of lack of heat, but also because of moisture conditions that are unfavorable for forest vegetation. In these zones, precipitation is significantly greater than potential evaporation, ensuring a high water content in the soil all year round. For this reason, the oxygen content of the soil is too scanty to permit the growth of normal tree roots.

The shift of the southern border of tundra to higher latitudes where the climate is less humid can be a fine example of the impact of moisture on the distribution of forest vegetation. This regularity can also clearly explain the absence of forests in regions of alpine meadows where the temperature sums are quite sufficient for the development of forest vegetation.

The next group of climatic zones with a moist climate (2) comprises three zones, as does the previous group.

Conditions of a moist climate with a moderately warm summer (2.3) are typical of a vast area of coniferous forest covering the north of the European USSR (except for the northern coastal regions) and most of the Asian territory of the Soviet Union. In this zone podzolic soils are well developed and coniferous forests consisting of evergreen and deciduous trees grow. The deciduous cone-bearing (larch) forests cover mainly the areas with the most severe winters. Although mean precipitation in the territory of coniferous forest may vary over a wide range, a significant water content in the soil and a rather high runoff coefficient are nevertheless observed all over the zone. That is why the coniferous forest zone is the place where many large rivers are formed.

The climatic zone with a moist climate and a warm summer (2.4) is situated in the middle of the European USSR and in the southern part of the Priamurje and Primorje territories. In this region mixed and broadleaf forest with many species of vegetation grow on turf-podzol and grey-forest soils.

A relatively small portion of the USSR territory is occupied by a belt of moist climate with a very warm summer (2.5), which embraces western Transcaucasus and the south of eastern Transcaucasus. Yellow and red soils are typical of this region. The broad-leaved forest there is to a great extent of a subtropical nature. Abundant precipitation, greatly exceeding possible evaporation, produces a comparatively high water content in the soil and ensures a high rate of runoff.

Forest vegetation is characteristic of climatic zones with a moist climate. The basic factor in favor of the dominance of forest landscapes in these areas is a high soil moisture content during the warm season, which at the same time does not exceed the limit beyond which a deficit of soil air prevents the development of the root systems of trees.

We should mention here that forest vegetation can exist only when there is significant transpiration, since turbulent diffusion is far more intensive at the level of the tree crowns than near the earth's surface. The enhanced turbulent exchange and the large total area of the leaves ensure a relatively high level of assimilation, allowing large nonproductive organs (trunks and branches) to develop. When soils are insufficiently moist and the transpiration necessary for trees is reduced, forest vegetation is replaced by steppe species that are more adaptable to soil moisture deficiency.

The insufficient moisture belt with a moderately warm summer (3.3) corresponds to the climatic conditions of some mountainous regions of central Asia and the south of central Siberia. This belt also includes a vast territory in the middle reaches of the Lena River and the lower reaches of the Vilyui River. The mountains of central Asia exhibit mountain steppe and sparse highland vegetation. Turfy-grassy and tansy steppes are typical of southern Siberia. In southern Siberia a part of this climatic belt is occupied by pine forests that are known to be of an intrazonal nature. Vegetation covering the area along the middle reaches of the Lena River is quite rich and characterized by coniferous forests in combination with forest-steppe plants.

The major portion of the insufficient moisture belt with a warm summer (3.4) occupies the south of the European USSR and western Siberia. A similar climate is also observed in some mountainous regions of the Caucasus and central Asia with diverse steppe and forest-steppe vegetation.

The insufficient moisture zone with a very warm summer (3.5) comprises relatively small southern coastal areas of the Crimea and the central Transcaucasus, a part of the eastern coast of the Caucasus, and the foothills of central Asia. In this zone, subtropical steppe, xerophytic forest, and other types of xerophytic vegetation are distributed.

All the zones of insufficient moisture conditions are characterized by a relatively small runoff coefficient, which generally is no more than 0.10–

0.20. Since the total annual precipitation is less than the potential evaporation, most of the precipitation is used up by evaporation.

Climatic zones with an arid climate occupy the greater portion of central Asia and eastern Transcaucasus.

The arid climate zone with a moderately warm summer (4.3) is situated in the Pamirs and corresponds to the conditions of highland deserts.

The arid climate belt with a warm summer (4.4) embraces semideserts and deserts where Artemisia and solonchak vegetation dominate.

Dry climate with a very warm summer (4.5) mainly covers desert regions with saxaul, shrub, and ephemeral vegetation. Light-chestnut, brown-desert-steppe, grey-brown desert soils, grey earths, takyr, and other soils typical of arid conditions are present in all these zones. The poor precipitation there is almost entirely used by evaporation; runoff usually approaches zero.

In climatic zones with insufficient moisture conditions and in zones of arid climate, the development of vegetation depends greatly on the soil moisture content. The reduction of soil moisture with an increase in climate aridity is accompanied by a decrease in the possible transpiration from a unit area. The nature of the vegetation cover also changes. It includes in the main different xerophytes and is no more dense.

Thus the main climatic zones are in accordance with general geographic zonality and particularly with geobotanic zonality.

Relying on the data presented in Table 6.6, we can conclude that during the warming trend associated with an increase in the temperature sums the geographical zones of surplus moisture and humid climate will shift to higher latitudes. The problem of the arid zones during a warming trend is more complicated since a change in their location depends essentially on fluctuations in the potential evaporation–precipitation ratio.

The above-stated data on possible changes in climate at the beginning of the next century show that the rates of precipitation and potential evaporation will increase over the greater part of the USSR territory and therefore their ratio will not change much. One might think that for this reason the position of geographical zones in the belt of insufficient moisture conditions will not change considerably. The boundaries of humid climatic zones will shift to a far greater degree.

Considering the possible variations in the air temperature sums during a warming trend and the dependence of the geographical zone location on the air temperature sums over the growing season, one can calculate the change in the zone location in response to climatic fluctuations.

The results of our calculation for the latitudinal belt 40–70°N are presented in Fig. 6.13: Part (a) shows the change in the average latitudinal boundary of the geographical zones in the Northern Hemisphere during

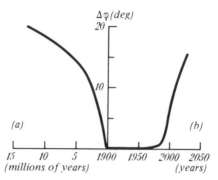

Fig. 6.13   Change in the mean latitude boundaries of geographical zones (a) in the Neogene and (b) at the present epoch.

the Neogene; part (b) depicts the same factor for the 20th and 21st centuries. It can be seen that at the end of the Tertiary the geographic zones of a humid climate moved southward, first at a slow rate and then more and more quickly.

This conclusion is confirmed by paleogeographic data that show that the natural conditions of the Neogene differed sharply from present conditions.

During the Miocene the central parts of western Europe were occupied by forests of evergreen plants including palms. In the north of Europe, Spitsbergen included, there were rich coniferous and deciduous forests comprising birch trees, beeches, oaks, pines, fir trees, and other plants. In the present steppe zone of southeastern Europe beech and oak groves expanded, including also some evergreen trees. Later, savanna vegetation developed here. Coniferous and deciduous forest covered the northern part of Asia.

In the Pliocene the major geobotanic zones shifted southward, though at that time they occupied higher latitudes than during the present epoch (Sinitsyn, 1965).

Figure 6.13b shows that the geographical zones would shift by 15° to higher latitudes before the end of the first quarter of the 21st century.

As mentioned above, the geographic zones in regions of insufficient humidity change their location to a far lesser degree, since the growth of potential evaporation due to the warming trend will be compensated for by an increase in precipitation. But even in this case the geographic zonality of the 21st century will differ greatly from that of the last centuries.

It must be mentioned that a change in the components of the natural environment cannot immediately follow a rapid climatic change. The change in the surface water regime will follow the climatic change closely.

Changes in vegetation cover, and especially in the soil, will be considerably slower.

Obviously all the components of the geographic environment can adapt only partialy to new climatic conditions within such a short period as a few decades. At the same time we should not underestimate the flexibility of many of these components to respond quickly to global climatic change. For instance, Grigorjev (1956) pointed out that during the relatively short-term warming of the 1920–1930s the northern margins of forests in some forest-tundra regions moved further into the higher latitudes.

It is obvious that the problem of changes in the position of geographic zones with a possible rapid change in climate calls for a more detailed treatment.

Productivity of vegetation cover

As mentioned above, the productivity of autotrophic vegetation cover is of great importance in the evaluation of the energy resources that support almost the whole mass of organic matter in living things of the biosphere.

Laboratory experiments and field research show that with sufficient moisture and other favorable conditions the productivity of autotrophic plants essentially depends on the atmospheric CO_2 concentration, which can range up to several times the present-day value. A quantitative estimate of the effects of CO_2 on the productivity can be obtained by means of numerical models of photosynthesis in the vegetation layer.

Many laboratory investigations (for example, Rabinovich, 1951) treated the dependence of the photosynthetic rate on the CO_2 concentration. They have established that, if the CO_2 content is not high, the photosynthetic rate increases almost linearly with the CO_2 growth. Then it slows down and stops increasing (the state of saturation) at a definite CO_2 concentration. With high concentrations of CO_2 the photosynthetic rate decreases and, after the CO_2 concentration reaches a certain level, photosynthesis can stop entirely.

The CO_2 concentration at which the state of saturation is reached can vary widely both for different plants and for the same plant under different environmental conditions.

Laboratory research shows that if light is intense enough and if sufficient heat, water, and necessary mineral substances are supplied, photosynthesis in many plants reaches a maximum value at a CO_2 concentration of several tenths of a percent. The application of carbonaceous fertilizers, which leads to an increase in the CO_2 concentration in the air of greenhouses, is based on this observation. Experimental data show that under such conditions it is possible to increase the yield by 50–100% or even more.

Under natural conditions the productivity of vegetation cover may be more dependent on atmospheric CO_2 concentration than would be expected from laboratory results obtained for individual leaves or isolated plants. This is explained by a considerably reduced CO_2 concentration in the vegetation layer under intensive photosynthesis as compared to the CO_2 concentration in the free atmosphere. The CO_2 concentration in the vegetation can approach the level of a "compensating point" in some parts of the vegetation layer when the photosynthetic rate is close to the rate of organic matter consumption for respiration and the productivity is zero.

For elucidating the vegetation productivity dependence on CO_2 concentration under natural conditions it is important to use numerical models of photosynthesis in the vegetation layer that take into account CO_2 variations with altitude. The models developed in a number of studies (Budyko, 1964; Budyko and Gandin, 1964, 1965, 1966) are based on the following concepts.

The dependence of the photosynthetic rate A for an individual leaf of a plant on photosynthetically active radiation Q and CO_2 concentration c can be written in the form

$$A = \frac{\beta Q}{1 + (\beta Q / \tau c)},$$ (6.1)

where β and τ are empirical parameters.

The vertical flux of CO_2 in the vegetation layer at a level z is

$$A_z = \rho k \, dc/dz,$$ (6.2)

where ρ is the air density and k the turbulent exchange coefficient.

The change of this flux in the layer dz caused by assimilation of CO_2 by leaves is

$$dA_1 = \frac{\beta Q s \, dz}{1 + (\beta Q / \tau c)},$$ (6.3)

where s is the surface of the leaves in a unit volume.

The change of the flux A_2 resulting from respiration of the leaves can be given in the form

$$dA_2 = -\epsilon s \, dz,$$ (6.4)

where ϵ is an empirical coefficient.

The general change in the flux of CO_2 is represented by the relation

$$dA_z = -dA_1 + dA_2.$$ (6.5)

From these formulas we can obtain a differential equation

$$\frac{d}{dz}\left(\rho k \frac{dc}{dz}\right) = -\frac{\beta s Q}{1 + (\beta Q/\tau c)} - \epsilon s,$$ (6.6)

where the function $Q(z)$ should be found from empirical data.

The boundary conditions for the last equation are

$$(\rho k \, dc/dz)|_{z=0} = -A_0,$$ (6.7)

$$(\rho k \, dc/dz)|_{z=H} = \rho D_H(c_\infty - c_H),$$ (6.8)

where A_0 is the CO_2 flux from the soil; c_∞ the atmospheric CO_2 concentration at the z_1 level, where it has little dependence on the properties of vegetation cover; c_H the CO_2 concentration at vegetation cover level H, and D_H an integral coefficient of turbulent diffusion in the air layer between H and z_1.

From the equation we can find A_H, i.e., the vertical CO_2 flux at the level H. Then the total assimilation A is determined by the formula

$$A = A_H + A_0 + \int_0^H \epsilon s \, dz.$$ (6.9)

Application of the above-stated method enables us to investigate the regularities of photosynthesis in the layer of vegetation cover, which is of special importance in elucidating the problem we treat, that is, the dependence of photosynthesis on the CO_2 concentration in the free atmosphere. As a result of the study carried out by Menzhulin, the greatest photosynthetic rate (i.e., saturation) for wheat occurs at concentrations of CO_2 somewhat above 0.2%. With a comparatively small increase in CO_2 concentration the relative increase of the photosynthetic rate will make up about half of the corresponding relative CO_2 increase.

It should be borne in mind that this conclusion refers to vegetation cover under conditions of sufficient light, heat, moisture, and mineral nutrition. Under less favorable conditions the dependence of photosynthesis on the CO_2 increase is weaker.

Of great interest is the estimation of this regularity for the entire biosphere at present-day rates of CO_2 growth. For some reason it is difficult to evaluate this regularity. Keeling and Bacastow (1977) think that, with small variations in the CO_2 concentration, the relative increase in productivity is 0.27 of the relative change in CO_2 concentration. Since most of the natural vegetation cover exists under conditions that are far from entirely favorable, their estimate does not contradict the result obtained by Menzhulin.

It must be mentioned that the conclusion of the Dahlem Workshop on the consequences of fossil fuel combustion (Lerman *et al.*, 1977) contains an evident error concerning the effects of anthropogenic CO_2 on photosynthesis. It suggests that the additional photosynthesis can consume only a portion of the annual CO_2 income produced by man's economic activity. Since that income is small compared to the photosynthetic comsumption of CO_2, the authors have concluded that the relevant photosynthetic increase will be also insignificant.

Actually photosynthesis does not directly depend on the rate of CO_2 release in the atmosphere but is determined only by the level of CO_2 concentration. If, for example, this level increases by 10% due to economic activities, the latter magnitude (not the anthropogenic CO_2 release into the atmosphere) is the upper limit of the relative photosynthesis growth under the influence of CO_2 increase.

Thus, other things being equal, an increase in the CO_2 concentration in the atmosphere is responsible for both an enhanced productivity of natural vegetation and higher grain yields. The productivity seemed to increase more in the second case than in the first.

Taking the above estimate of Menzhulin as the first approximation and assuming that the CO_2 concentration has increased by 10% since the end of the 19th century, it is possible that the latter change is responsible for the increase in wheat yield (and probably that of other crops) of about 5%. The actual increase in the average wheat yield over the indicated period of time was far greater than this value, but it is impossible to verify our estimate by data on changes in yields because of the stronger impact on yields of progress in agrotechnics and of some other factors. Nevertheless, the CO_2 effects mentioned here can be of essential importance in the economy.

We should emphasize that in some regions an increase in CO_2 concentration favorable for plant productivity could be more or less compensated for by a worsening of climatic conditions. As mentioned in the previous section, during a warming trend, average precipitation sums decrease and droughts are more frequent over part of the continental areas in the middle latitudes of the Northern Hemisphere. It is rather difficult to estimate quantitatively the effects of such climatic change on average yields. Undoubtedly, over areas subjected to a severe drought, the adverse weather would affect crop yield more than the CO_2 increase, whose effect in this case is less significant. However, if we assume that as a result of climatic change droughts that decrease the crop yield by 15% will occur 10–30% more often and that during years of sufficient precipitation the yield will be about 10% higher due to the increase in CO_2, the rise in the total crop yields for approximately a decade could appreciably outweigh the damage caused by droughts.

Figure 6.14 shows the dependence of wheat productivity on CO_2 concentration and precipitation. The results are based on Menzhulin's (1976) model, constructed for the average climatic conditions of wheat cultivation in the USSR. The graph shows that the productivity grows proportionally to both the CO_2 concentration and the amount of precipitation. A 10% increase in CO_2 or precipitation leads in either case to an increase in the productivity of 4–5%.

Since a similar dependence should also reveal itself in the case of natural vegetation under conditions of sufficient humidity, we think that the change in the total productivity of autotrophic plants is in accordance with the relevant changes in the atmospheric CO_2. Variations in the thermal regime of the earth's surface that are associated with fluctuations in CO_2 add to the productivity change.

If all other conditions remain the same, an increase in the temperature of plants often leads to a decrease in the productivity due to an increase of organic matter expenditure for respiration. But a global warming trend contributes to an increase in the total productivity of continental vegetation, due mainly to expansion of the areas occupied by more productive plants.

We might suppose that the global productivity of vegetation with an increase in atmospheric CO_2 concentration must become higher since it is

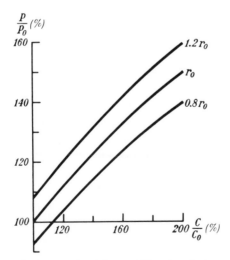

Fig. 6.14   Dependence of wheat productivity on CO_2 concentration and the amount of precipitation.

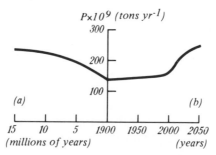

Fig. 6.15   Change in plant productivity (a) during the Neogene and (b) at the present epoch.

directly affected by CO_2 variations, by an increase in the total precipitation over some part of the continents, and by an expansion of the more
productive zones to higher latitudes. It is very difficult to take all these
factors into consideration. Some rough estimates concerning these relationships, obtained by Efimova, are presented in Fig. 6.15. Part (a) shows
the variation in the total productivity of continental vegetation in the late
Neogene. Part (b) depicts a possible change in future productivity with
continuation of global warming. It is quite possible that the 50% increase
in total productivity that is shown in the figure will not actually occur in a
few decades, since the productivity of the natural vegetation could not
reach the maximum corresponding to the rapidly changing environment in
such a short period of time.

However, we expect that the average productivity of both natural vegetation and arable lands will increase considerably as the CO_2 concentration goes up.

Practical problems

Chapter 2 contains data showing that over the last hundred million
years the CO_2 concentration in the atmosphere tended to decrease. As a
result, the climate became cooler and polar glaciations appeared, repeatedly spreading to the middle latitudes in the Quaternary.

The decrease in CO_2 led to a reduced rate of photosynthesis in autotrophic plants, which lessened the total mass of organic matter in the biosphere.

A further decrease in CO_2 concentration could threaten life on the earth
and destroy the biosphere within a relatively short time, from the geological standpoint.

By using the coal and oil reserves accumulated for hundreds of millions
of years, man is rapidly restoring the chemical composition of the ancient

atmosphere. It follows from the above estimate that it will take only a few decades to reach the level of CO_2 that corresponds to the late Tertiary period, i.e., several million years ago.

The process going on now can be considered as "rejuvenation" of the biosphere, i.e., the restoration of natural conditions with a warm climate at all latitudes, without polar ice cover, and with high productivity of autotrophic plants, which can sustain a far greater mass of heterotrophic living organisms as compared to the modern epoch.

This change in natural conditions could be especially favorable for countries in zones with a cold continental climate where the development of agriculture and many other branches of economic activity are difficult.

If a return to the climatic conditions of the Tertiary occurred rather slowly, this process could be favorable for mankind. But the enormous rate of this process creates a number of problems whose solutions cannot be so easy.

An opinion has been repeatedly expressed that any sudden change in climatic conditions embracing vast areas leads to severe economic consequences involving large investments in reorganization of all the branches of economic activity dependent on climate. At the same time, inadvertent climatic change over extensive territories evidently would not be helpful for man's economic activities in all cases, since in some regions such change would more or less hamper these activities.

The impact of the warming trend on agricultural productivity will evidently be of most practical importance. As mentioned above, this impact will differ in various climatic zones. The warming trend will raise the productivity of crops in regions with a comparatively cold and humid climate where a longer growing season and increased temperature sums over this season could result in a considerably higher crop yield. At the same time, during certain periods of the warming trend, precipitation could decrease in a number of regions with unstable moisture conditions. This would have an adverse effect on productivity. Since in any case the CO_2 growth will contribute to the yield increase, it might be thought that the general impact of the warming trend on productivity will be positive on a global scale.

One of the most important results of the expected climatic change is associated with the river runoff rate, since fresh water resources even now in a number of regions do not satisfy existing demands. It follows from the previous section that by the end of the 20th century moisture conditions could worsen in some territories, but somewhat later, during the Arctic warming, river runoff will increase considerably. This increase in water resources will occur, however, in zones of sufficiently humid climate.

Among other important practical problems associated with the warming effects we should mention the melting of the surface layers of permafrost soils which cover almost half of the USSR territory. Some difficulties could arise there concerning the preservation of buildings in these regions. The possibility of a contraction of interior continental glaciers, some of which are essential for the water regime of arid regions, could also be of great importance.

Grosswald and Kotlyakov (1978) have treated the question of the impact of global warming on glaciers and have concluded that all the Arctic glaciers, except those of Greenland, could disappear rather quickly. They also think that the mountain glaciers in middle latitudes will response more slowly to the warming.

Possible changes in the world ocean level associated with the glaciation regime are also noteworthy.

As we know, over the last hundreds of thousands of years the world ocean level repeatedly rose or descended according to the development of land glaciers. During the greatest Quaternary glaciations the ocean level decreased by over 100 m compared to its present-day position since a considerable amount of water vapor transferred from the ocean to the continents was used in the formation of ice cover. During warm interglacial epochs the ocean level was up to several tens of meters higher than at present. The complete melting of the present glaciers could cause a 66-m rise in the ocean level. It must be mentioned that the Antarctic glaciers and the Greenland ice sheets comprise almost the whole water mass of present-day glaciers. (The melting of the former would increase the ocean level by 59 m; that of the latter by 7 m). The 66-m ocean level rise would result in an ~3% reduction in total land area. The magnitude of this area seems to be comparatively small but the land involved includes a number of areas where the largest cities of the world are situated.

Possible consequences of the partial melting of present-day glaciers, which would lead to an ocean level rise of a few meters, are very important. We can conclude that such a rise would be disastrous for nations situated in lowlands, such as the Netherlands, and for some large coastal cities. It should be borne in mind that even coastal towns that are situated several meters above see level would also suffer greatly, for their port facilities would be flooded. Therefore, even a relatively insignificant rise in the ocean level will present global problems.

The question of the fate of the Antarctic and Greenland glaciers during the warming trend has been discussed only recently.

The first work in this field was presented by Mercer (1978), who concluded that the western part of the Antarctic glacier will melt comparatively soon (perhaps even in 50 yr) as a result of the forthcoming warming trend. After this event the ocean level will rise by ~5 m.

This possibility arises from the specific nature of the western Antarctic glacier, which rests on the sea bottom and not on land, as does the eastern part. Such glaciers are very sensitive to fluctuations in sea water temperature and can easily succumb to a temperature rise. Grosswald and Kotlyakov (1978) have made a conclusion similar to Mercer's on the basis of a more detailed treatment of the problem. According to Thomas, Sanderson, and Rous, the destruction of this glacier as a result of the warming trend could take a few hundred years. At the same time they do not exclude the possibility of a faster development of this process (Thomas *et al.*, 1979). The necessity for a detailed treatment of this vital issue is quite obvious.

One of the most acute problems that will arise with a rapid climatic change is associated with the task of preserving many species of wild animals and plants. The destruction of specific ecological systems will threaten their existence. The question of the preservation of food fish reserves, which are sometimes very sensitive to climatic changes, is a part of the same problem. There are a great many analogous problems in this field because a global climatic change will affect many components of the environment.

We can obtain a general estimate of the impact of global warming on economic activities only through extensive interdisiplinary researches, which have not yet been carried out. Since many aspects of this impact will be unfavorable, it is quite natural to ask if it is possible to prevent the global warming trend either entirely or partly.

Such an opportunity may arise in the more distant future when new methods of energy production may be developed that will enable us to control the energy balance and preserve favorable climatic conditions. At the same time, according to the available prognoses for the nearest decades, it is hardly possible to avoid using large amounts of fossil fuels.

A method for modifying climatic conditions that could help in averting the globel warming trend was set forth in Budyko (1974). Chapter 4 presents estimates of stratospheric aerosol effects on the shortwave radiation flux reaching the earth's surface. These estimates show that relatively small fluctuations in the amount of aerosols in the lower stratosphere can essentially affect the "meteorological solar constant" and accordingly the mean air temperature near the earth's surface.

Let us consider the possibility of changing the climate by increasing the aerosol content of the lower stratosphere. For this purpose we shall estimate the aerosol mass that must be released into the stratosphere to induce a climatic change of the same scale as the 1920–1930s warming but one that will cause a reduction in the mean temperature near the earth's surface.

The data of Chapter 4 show that in this case it is necessary to decrease the mean value of the direct solar radiation by 2%, which corresponds to a 0.3% decrease in the total radiation.

According to the above data, we assume that the mean aerosol content that will reduce the total radiation by 1% is 0.8×10^{-6} g cm^{-2} and we find that for our purpose 600,000 tons must be added to the aerosol mass in the Northern Hemisphere.

Such an increase in the aerosol mass can be achieved by different technological means. The method we shall consider here is based on the sulfate nature of the stratospheric aerosols.

In order to increase the concentration of sulfate particles it is sufficient to introduce into the aerosol layer a certain amount of sulfur dioxide, which then turns into droplets of sulfuric acid. About 400,000 tons of SO_2, which can be produced by burning 200,000 tons of sulfur, are required for the formation of 600,000 tons of sulfuric acid.

Let us assume that the mean residence time of the stratospheric aerosols is 2 yr. To maintain the necessary concentration of aerosol particles consisting of pure sulfuric acid, 100,000 tons of sulfur must be released into the lower stratosphere each year. However, because of the hydroscopic properties of concentrated sulfuric acid, the droplets absorb water vapor from the air, thereby increasing their mass and decreasing their specific weight. As a result, we shall need a considerably smaller amount of the reagent for a given reduction of the total radiation. Since we do not know the exact sulfuric acid concentration in the droplets of the stratospheric aerosols, we shall assume that it is $\sim 50\%$. In this case only 40,000 tons of sulfur per year are required to perform our task. We note here that this amount is less than half of the above-mentioned mass of sulphur because of the reduced specific weight of the aerosol when sulfuric acid is diluted with water.

Taking into consideration the fact that modern high-altitude aircraft can carry a load of about 15 tons to the level of the aerosol layer, we find that the necessary amount of reagent can be transported to the lower stratosphere by several regular aircraft equipped with an arrangement for burning sulfur in the atmosphere.

It is interesting that the amount of reagent to be introduced into the atmosphere according to the above method for altering the climate is an infinitesimal fraction of the material that is released into the atmosphere as a result of modern man's economic activities. The proposed amount is tens of thousands of times less than the mass produced by human activity, which according to current data is hundreds of millions of tons a year (SMIC, 1971).

The amount of reagent that would settle to the earth's surface eventually is also absolutely insignificant, i.e., at the indicated rate of use, it would average about 0.2 mg S m^{-2} yr^{-1}, which is about a thousand times smaller than the natural fallout of sulfur with atmospheric precipitation (Budyko, 1974). It is obvious that such quantities are of no importance in the pollution of the environment.

It should be mentioned that although the calculations that are the basis for the above conclusions are fairly approximate, we believe that more accurate calculations of subsequent investigations will confirm that, in principle, it is possible to change the climate by modern technological methods by controling the aerosol concentration in the stratosphere.

There are suggestions for averting the warming trend by direct influence on the CO_2 balance.

For instance, Marchetti (1976) and Nordhaus (1977) have introduced a proposal for removing CO_2 from the waste products of fossil fuel combustion by converting it to the liquid state under high pressure and pumping it into deep ocean layers. The technological and economic possibility of the realization of this plan are at present obscure.

We think that the technological realization of any methods for changing the global CO_2 balance or the climate itself is not a matter of current interest, for the application of these methods will yield complicated and contradictory results, which could be unfavorable for individual countries. Therefore, it is unlikely that such methods will be used in the near future.

For this reason it is necessary to study the practical consequences of the expected climatic changes due to a warming trend in order to consider beforehand the means of using most profitably the favorable results of such a change and averting the adverse consequences to the greatest possible extent.

6.3 THE DISTANT FUTURE OF THE BIOSPHERE

The natural evolution of climate

The problem of the distant future of the biosphere is only of theoretical interest, unlike that of the climatic variations and changes in the biosphere over the next few decades, which are exceedingly important for man's activities.

Scientists interested in this field ran into many more difficulties than those concerned with the immediate prospects for the development of the biosphere. This restricts the possibilities of obtaining reliable information on the distant future of the biosphere.

In considering this problem we shall dwell on the natural climatic variations that occur over long intervals of time. These variations can depend on the following causes:

1. *Astronomical factors.* A change in the position of the earth's surface relative to the sun results in a climatic change with a time scale of tens of thousands of years.

2. *Composition of the atmosphere.* During Tertiary and Quaternary time the climate was definitely affected by a decreasing CO_2 concentration in the atmosphere. Considering the rate of the CO_2 decrease and the corresponding change in the air temperature, we can conclude that the effect of natural changes in CO_2 on the climate is important for time intervals over one hundred thousand years.

3. *The structure of the earth's surface.* A change in the surface relief and associated alterations of the coastlines can appreciably influence climatic conditions over large expanses during the periods of at least hundreds of thousands or millions of years.

4. *The solar constant.* Possible variations in solar radiation due to the sun's evolution should also be taken into consideration. These variations can strongly affect climatic conditions over periods of more than a hundred million years.

The temporal scales of the above-mentioned effects on climatic conditions agree with similar estimates by Mitchell (1968) and other authors.

Let us now look at the effects that these factors can produce on climatic conditions.

Future changes in the position of the earth's surface relative to the sun can be estimated quite precisely. Calculations in this field (Sharaf and Budnikova, 1969; Vernekar, 1972) allow us to evaluate the fluctuations in solar radiation income during the warm half-year, which are mostly responsible for changes in the polar ice boundary. According to Milankovich (1930), radiation fluctuations at the latitudes close to the polar circles produce a particularly strong effect on the ice shifts.

The available calculations show that in 10 to 15 thousand years the amount of radiation will drop noticeably in the region of the "critical latitudes" of the Northern Hemisphere, the decrease being about two thirds of the reduction that occurred during the last Würm glaciation. This drop could lead to the onset of a new ice age with the development of continental glaciations that, would, however, be somewhat less extensive than the Würm glaciations.

Much greater decreases in radiation in the critical latitudes of the Northern Hemisphere could occur in hundreds of thousands of years; this would mean the development of great glaciations at that time.

The influence of astronomical factors on climate can intensify with a further decrease in the CO_2 content of the atmosphere. Since for many millions of years the CO_2 concentration in the air tended to decrease, in the future it might go on decreasing in the absense of man's impact on the climate. The data on the CO_2 decrease at the end of the Cenozoic presented in Chapter 4 enable us to obtain an approximate rate for the expected decrease.

Extrapolating these data we can estimate how long it would take for the atmospheric CO_2 concentration to decrease from the present-day level of $\sim 0.03\%$ to the level at which complete glaciation of the planet is possible. According to the semiempirical theory of the thermal regime, this value is $\sim 0.015\%$. The data on the rate of CO_2 decrease in the late Cenozoic show that the difference between the above estimates approaches the change in CO_2 concentration since the beginning of the Quaternary glaciations. Therefore, we can conclude that the complete glaciation of the earth could occur over a period of the order of a million years.

In addition to the general tendency toward cooling induced by a decrease in CO_2 concentration, in the presence of polar ice there are periods when the ice cover tends to expand as a result of a change in the position of the earth's surface relative to the sun. Taking this fact into consideration, we must decrease our estimate of the time at which complete glaciation of the earth could occur.

In view of the above ideas we can draw the following picture of the possible evolution of the biosphere.

The continuing decrease in CO_2 concentration will be accompanied by a gradual reduction in the productivity of autotrophic plants and in the total mass of living organisms on the earth. At the same time the zone of polar glaciations will expand and move to lower latitudes with the coming of ice epochs.

In a few hundred thousand years or in a few million years the ice cover will reach a critical latitude and then expand to the equator in a self-propelled process. As a result, the planet will be completely covered with ice, which will be stable because of a low negative temperature at all latitudes.

Probably all biological processes on the earth will cease under the conditions of complete glaciation. This supposition is based on the fact that during the long period of the Antarctic glacier's existence no living organisms have appeared that could live constantly in the central parts of that glacier. With the expansion of central Antarctic conditions in a relatively brief period to the entire globe, we can hardly expect living organisms to evolve and adapt to such unfavorable conditions.

Complete glaciation of the globe corresponding to the thermal regime designated by point B' in Figs. 3.17 and 3.18 would be very stable because at all latitudes the air temperature will be far below the freezing point of water. It is also possible that in the distant future this glaciation will disappear as a result of a gradual increase in the solar constant. As mentioned in Chapter 4, some investigations predict that with the evolution of the sun its surface temperature will increase and the amount of emitted energy will grow (Schwarzschield, 1958; Ångström, 1965; White, 1967).

It is seen from Figs. 3.17 and 3.18 that for the melting of planetary glaciation the solar constant must increase by ~40%. According to the data in the above studies this might occur in a few billion years. In this case ice will melt at all latitudes and the air temperature will rise by more than 100°, reaching an average value of about 80°C. Since with a further increase in the solar constant the temperature of the earth's surface would continue to rise, the melting of the ice would hardly create favorable conditions for life to originate again on the earth.

Realizing that our knowledge of the sun's evolution is rather vague, we emphasize the hypothetical nature of the above change in the earth's temperature regime. However, if such a change did occur, the thermal regime would develop according to the main part of the hysteresis curve in Figs. 3.17 and 3.18, i.e., from the ice-free regimes of Mesozoic time (somewhere above E) to modern conditions (3) and from complete glaciation (B' to A) to its disappearance (A').

The above estimation of the time during which complete glaciation of the planet could occur is very rough for a number of reasons.

We think that the result obtained is interesting mainly as an illustration of a specific property of the Quaternary glaciations mentioned in our previous work (Budyko, 1971). It is possible that, unlike the Permo-Carboniferous and other ancient glaciations, the Quaternary ones were not temporary episodes in the earth's evolution but a transitional period from stable ice-free climatic conditions to a far more stable regime of complete glaciation of the planet. This transitional period, during which, according to the above estimate, the biosphere could have ceased to exist, is fairly brief compared to the entire period of the existence of life on the earth.

Complete glaciation of the planet in the course of the natural evolution of the earth's climate seems more probable in view of the present-day tendency toward an uplifting of the continental level in some high and middle latitudes of the Northern Hemisphere.

According to paleogeographic maps by Saks (1960) and other authors, at the end of the Jurassic the ocean covered almost half of the latitudinal belt at 70°N and ocean waters could easily circulate between the middle and polar latitudes. In the early Cretaceous the area occupied by the continents at this latitude began to increase slowly, a process that continues at present. As a result, the continents now occupy $\frac{5}{8}$ of this latitudinal belt. Therefore, the straits connecting the seas of the middle and high latitudes of the Northern Hemisphere are $\frac{1}{3}$ of their former width. As a result of this, the meridional transfer of heat to the polar latitudes by sea currents decreased considerably, creating the prerequisites for the development of the Arctic glaciations.

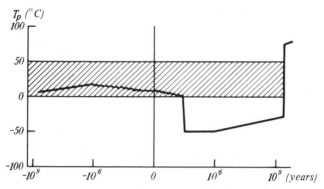

Fig. 6.16   Change in mean air temperature at the earth's surface.

There is no reason to suppose that an intensifying tendency toward isolation of the Arctic polar basin, which has existed for many millions of years, will stop in the near future. Continuation of this tendency means that in the future new advances of glaciers and their expansion to the critical latitudes are even more likely.

The effects of changes in the earth's relief on the future climate may reveal themselves later than those of decreasing CO_2 concentration in the atmosphere and therefore they may be insignificant. However, since these estimates are very approximate, the question of the comparative importance of both factors to future climatic conditions seems to be unsolved.

A hypothetical diagram of variations in the mean global air temperature near the earth's surface in the future that is based on the above concepts is presented in Fig. 6.16, where variations in the mean global temperature in the past can be seen as well.

To make our example clearer, the time scale for the range of one million years in the past and in the future is a thousand times greater than that for more distant times. The range of air temperature variations that comprises the most favorable conditions for the development of biological processes is shaded. It must be emphasized that the diagram is highly hypothetical, since it is based on a number of the assumptions mentioned above.

The biosphere of the earth and life in the universe

In order to understand the prospects of the far distant future of the earth's biosphere it is important to study the problem of the existence of other biospheres in the universe. For many centuries attempts have been made to solve this problem.

As we know, many of the philosophers of antiquity believed in the existence of life on various celestial bodies. That concept was widely accepted in the eighteenth and, particularly, in the nineteenth centuries among prominent scientists in the field of astronomy. The supporters of the concept thought that the inhabitants of the planets and stars were reasoning creatures and that there were numerous civilizations in the universe.

The launching of artificial earth satellites held out new prospects for studying the question of extraterrestrial life. During flights to the moon and Mars attempts were made to find living organisms there. But even before these flights, many scientific and popular books that were devoted to this question appeared, among them books by Shklovsky (1973) and Sagan (1975).

In addition to the investigations carried out to find life on other planets, the last few years have witnessed attempts to receive information that might be transmitted by representatives of extraterrestrial civilizations and to transmit information into space by means of radio. Several international conferences (one of them in the USSR, Byurakan, 1971) concerned with this matter were also held, indicating a rising interest in the question of extraterrestrial civilizations.

Some scientists are sceptical when treating this problem. For example, the famous biologist G. G. Simpson expressed the following view in 1963 before the implementation of the U.S. program to investigate the existence of life on the moon and Mars. Later it appeared in his collected articles:

In the case of extraterrestrial life, what real evidence there is pertains only to our solar system and it is strongly, although not quite conclusively, opposed to the presence of life on any other planet of this system. When scientists argue for the existence of such life, they are thus discussing a partly scientific question but are on the wrong side. Outside of our solar system there is absolutely no objective evidence at all, either pro or con. When scientists discuss that issue they are at best extrapolating probabilities beyond any presently testable point. At worst, and usually, they are simply fantasying (Simpson, 1969, pp. 73–74).

According to Simpson, it is practically impossible to obtain any information on the existence of life outside the solar system and there is no hope at all of discovering an extraterrestrial civilization.

Not only biologists, but also some specialists in the exact sciences concerned with space investigations are rather sceptical, regarding the search for extraterrestrial civilizations as useless. For instance, Sagan (1975) recalls the views on this problem expressed by Keldysh and by an American astronomer, both of whom doubt the fruitfulness of such research.

Since the direct methods used to study the question of extraterrestrial life have met with great difficulties, it seems appropriate to apply other approaches in this field. One of these may consist in investigating those regularities of the evolution of the earth's biosphere that can help in understanding life conditions in the universe. In such investigations it is necessary to answer questions concerning the factors that permitted life on the earth to exist for a tremendously long period of time.

Paleontological evidence shows that living organisms have continually existed on our planet for more than 3 billion years. This is a very long period, being over half of the time elapsed since the earth's origin and more than 10% of the time since the "big bang," when the universe began to form. Since there is no information on the distribution of life in outer space, we do not know what chances living things have for existence on a planet for such a long period of time. They could be either relatively great or quite small.

This problem involves some other aspects. What is the probability of a sufficiently long existence of the biosphere for reasoning creatures to originate, develop, and start investigating their environment by scientific methods? The fact that the time elapsed from the emergence of life on the earth until the development of modern civilization is great even on a cosmic time scale can be easily explained by the slowness of the progress of living organisms. Although the concept of progress as applied to an evolutionary process cannot always be easily interpreted, this term is sufficiently clear for discussing the question of human origins.

The reason for the slow progress is clear from the following remark:

In fact what natural selection favors is always and only the genetic characteristics of those organisms that leave the most descendants under existing conditions in any one period. If those leaving the most descendants are the fittest in some sense, for example the most intelligent, then natural selection does maintain or increase fitness in that special sense. But if . . . those less fit in one special sense, less intelligent, say, have more descendants, then natural selection will *decrease* fitness in that sense (Simpson, 1969, pp. 56–57).

These words do not contradict the well-known fact that in the course of evolution the tendency toward complication of living organisms prevails, the development of the nervous system in animals being a particular example. For instance, Jerison (1973) has established that during the history of vertebrates their brain size gradually increased relative to their body size. In particular, the data for mammals are presented in Table 6.7.

Note that in this case a considerable increase in the brain size occurred over several tens of millions of years. For the direct ancestors of man this process was quicker: the same increase in brain size took several million years.

With these and similar data in mind we can conclude that during the history of the biosphere conditions have been favorable not only for the preservation of living organisms but for their progressive development.

Table 6.7
Relative Brain Volume in Mammals[a]

	Herbivora	Predators
Early Tertiary primitive forms	0.18	0.47
Early Tertiary progressive forms	0.38	0.61
Mid and late Tertiary forms	0.63	0.76
Modern forms	0.95	1.10

[a] The mean value for contemporary animals is assumed to be equal to 1.0.

The question of how such conditions have been maintained is worth particular notice.

The biosphere's stability

In the author's works (Budyko, 1971, 1974, 1977a) it has been mentioned that, contrary to the ideas found in popular literature, living organisms have exploited only a very limited part of our planet. The total mass of living substance is of the order of $n \times 10^{18}$ g, which corresponds to a surface density on the earth of ~ 1 g cm^{-2}. Since wood, which has a very low metabolism rate, is the main component of the total biomass, it might be thought that the amount of biologically active living substance does not exceed 0.1 g cm^{-2}. This value is negligibly small compared to the mass of the atmosphere, hydrosphere, and upper layers of the lithosphere contained in a vertical column of 1 cm^2 section, i.e., the main components of the environment surrounding living organisms.

Assuming that the mean thickness of the earth layer where organisms exist is of the order of 10 km, it is evident that living substance (with an average specific weight close to 1) occupies not more than one millionth of the volume of this layer.

As mentioned earlier (Lavrenko, 1949), living substance could be considered as a very thin (in comparison to the planet size) film situated at the surface of the continents, in the soil, and in the upper layers of water basins. This film is partly broken: it vanishes completely or almost completely in the zones with permanent ice and snow cover, in the driest continental regions, and in the ocean regions with small amounts of the mineral matter necessary for plants. Thus, over billions of years of evolution, living organisms have not sufficiently adapted to conditions existing over a part of our planet's surface. To a still lesser degree have they adapted to the conditions of other regions of the biosphere: the quantity of organisms in the atmosphere, in the deep layers of the oceans, and in the lithosphere beyond the soil layer is as a rule negligibly small, even compared to their relatively small mass in the surface film.

It is worth noting that the areas occupied by individual species of animals and plants are usually very limited. These areas for the great majority of species cover only a small part of the earth's surface, and temporal fluctuations in the natural conditions within these zones frequently lead to a sharp change in the number of organisms, and sometimes to their extinction.

The boundaries of the life zone during the main part of history of the biosphere were to a considerable extent determined by the possibility of maintaining the photosynthesis of autotrophic plants. This can be attributed to the fact that almost the whole mass of living substance (particularly that of multicellular organisms) exists by the redistribution of the solar radiation energy consumed by autotrophic plants. The study of photosynthesis shows that this process takes place within a rather narrow temperature range and requires water resources accessible to the plants, definite CO_2 concentrations, and necessary mineral substances. A vital condition of photosynthesis is the absorption by plants of sufficient amounts of radiation in a definite range of wavelengths. The combination of conditions essential for photosynthesis does not occur over the entire surface of the earth and, over a considerable part of the surface, it is not observed in all seasons. The combination of natural conditions that provides high productivity of autotrophic plants and therefore results in the existence of a large mass of heterotrophic organisms per unit of area occupied is even more rare.

Although, in the present epoch, in a number of regions with a humid tropical climate, the productivity of the natural vegetation canopy reaches rather high values, it is much less than the productivity expected with a higher CO_2 concentration and a lower O_2 concentration in the atmosphere.

The question of the productivity of photosynthesis is of decisive importance in the explanation of the relatively small amount of living substance in the modern biosphere. The author's work (Budyko, 1971) deals with the study of the dependence among plant productivity, the biomass, and the factors influencing photosynthesis. Considering this dependence one can, in particular, establish that the total biomass depends appreciably on CO_2 concentrations in the atmosphere and in water basins. An increase in this concentration should lead to a growth in the total biomass.

In the above work the efficiency of use of the energy and water resources by the natural vegetation canopy has been estimated. For the earth as a whole plants consume only about 0.1% of the solar radiation coming to the earth's surface. Approximately the same fraction of precipitation is spent by land plants for the synthesis of organic matter.

Thus the life zone of organisms covers a small part of the range of natural conditions in the external envelope of our planet. The amount of living substance existing within this zone is negligibly small compared to the volume of its surroundings. Taking into account that the indicated range is very narrow in comparison with the changeability of the physical and chemical conditions on the planets of the solar system (to say nothing of the conditions on other planets of the universe), it might be thought that the flexibility of living organisms (i.e., their adaptability to various surroundings) is very limited compared to the changeability of these conditions even on our planet and especially in the universe as a whole.

In connection with the narrowness of the life zone and with the limited amount of living substance in the external envelope of the earth, in order to understand the reasons for the antiquity of the biosphere we should study the mechanisms that maintain its stability.

The stability of the biosphere depends on the regularities governing its development under more or less constant external conditions and on the variability of the external factors that influence it. The balanced character of the ecology of earth's biosphere is a typical and to some extent paradoxical feature. As is known, heterotrophic organisms are related to one another and to autotrophic organisms by trophic relations that frequently result in the annihilation of one organism by another. If predators, which obtain food most effectively by the extermination of pursued animals, leave a greater number of descendants, it would seem that natural selection further promotes the indicated ability of the predator. The same assumption can be made concerning the ability of herbivorous animals to consume the increasing amount of living plant matter. It is easy to see that similar consequences of natural selection would result in the quick destruction of the most prosperous species because they annihilate their sources of food.

Although ecological works have for a long time dealt with the study of the complex mechanisms preventing the possibility of mutual destruction of species in the course of trophic interrelations, it is believed that this study is still far from complete. The use of numerical models of ecological systems is essential for progress in this investigation. In these studies it has been established that the system of negative feedbacks that decrease the variability of the size of populations incorporated into ecological systems is very important for increasing the stability of ecological systems.

The high efficiency of these feedbacks is proved, for example, by the antiquity of many present species and genera, which have existed, in some cases, for tens and hundreds of millions of years. The fact that the process of evolution of organisms is very slow under constant external conditions is revealed by the existence of various "living fossils." For example, a great number of ancient forms have remained in the regions of the open oceans, where environmental conditions were most constant.

As noted earlier, ecological systems have definite features of integrity. For this reason they are sometimes considered as "superorganisms." A similar term can be applied to the global ecological system—the biosphere.

The use of this term is warranted by the fact that organisms exert a considerable influence on the many components of the biosphere and particularly on the atmosphere, where the main mass of oxygen is produced as a result of the activity of autotrophic plants, which are also of great importance for the CO_2 balance. Atmospheric oxygen, organic carbon in the atmosphere and oceans, and liquified and gaseous water in the biosphere have comparatively short exchange cycles due to the activity of living organisms. Sediments in the upper layers of the lithosphere are created, to a great extent by organisms.

It is worth noting that the effect of organisms on the environment can improve the conditions of their existence mainly in local ecological systems and for comparatively short periods of time. The results of this effect on the global ecological system apparently cannot have adaptive significance because the duration of most species is usually much shorter than the period of large scale variation of the biosphere.

It is characteristic that in some cases the environment is adversely affected by the activity of organisms: as mentioned above a high level of oxygen concentration produced by plants in the modern atmosphere lowers their productivity. The withdrawal of some atmospheric CO_2 in the course of its biotic cycle has led to a photosynthesis productivity decrease. Thus the effect of organisms on the biosphere does not increase its stability, even with constant external conditions.

It has been repeatedly proposed that great changes in the biosphere are associated with the effect of external factors. Such assumptions refer, in particular, to the critical epochs of geological history (such as the extinction of many groups of animals at the end of the Permian, Triassic, and Cretaceous), which are possibly the consequences of volcanic activity fluctuations. (This will be considered in more detail below.)

Among the examples of the evident influence of external factors on organisms is the extinction of numerous animal groups at the end of the Tertiary period in South America. There is no doubt that the joining of South America with North America that resulted from the land elevation in the region of the Isthmus of Panama was a cause of the animal extinctions. The penetration into South America of highly developed mammals led to the rapid disappearance of many representatives of the original fauna of South America that existed there during almost the entire Tertiary period in conditions of isolation from the other continents.

The balanced ecology of the biosphere has been appreciably disturbed as a result of man's activities, which, because of their specific character, can be considered an external factor relative to various components of the biosphere.

We observe that noticeable changes in the biosphere's ecology under the influence of anthropogenic factors had already appeared in the early stages of the existence of *Homo sapiens* and were manifested in the annihilation of many species of large animals by our ancient ancestors (Martin, 1967; Budyko, 1971).

Man's impact on animate nature increased sharply in the present epoch when, as a result of the inadvertent consequences of economic activities, many species of organisms have been annihilated and the populations of a great many more species have been drastically decreased. The ease with which man can annihilate living organisms over vast areas is one of the proofs of the high sensitivity of the biosphere to variations in the external factors that influence it.

The duration of the biosphere's existence in the presence of fluctuations in external factors affecting the global ecological system depends on its stability as determined by a set of direct relations and feedbacks between the system components and external factors.

Laying aside the above question of man's activities, consider the main external factors whose variations can affect the global ecological system. Among them are solar radiation and the exchange of matter between deep layers of the lithosphere and the biosphere.

The importance of the first of these factors is evident since, with the expiration of solar radiation (which could occur in several billion years), the biosphere will cease to exist.

The studies whose results are discussed in Chapter 3 have shown that climatic conditions can vary significantly with even relatively small variations in solar radiation.

A reduction of the solar constant by only a few percent could result in complete glaciation of the earth, which would be followed by a 60–100° lowering of the mean surface air temperature. Such a change in climate would, it seems, result in the destruction of all living organisms, i.e., in the disappearance of the biosphere.

The high sensitivity of the thermal regime to solar radiation fluctuations enables one to suppose that the existence of the biosphere depends on even relatively small variations in solar emission. Apparently this emission is rather stable. As mentioned in Chapter 4, numerous attempts undertaken to determine from observation how stable the "solar constant" (i.e., the solar radiation flux at the outer boundary of the atmosphere) is have led to the conclusion that over the last decades this value has hardly varied or has fluctuated within a narrow range, usually not exceeding 0.1% of its value.

Of great importance is the question of solar constant variations in the geological past, which was treated in Chapter 4. It is assumed that the evolution of the sun will lead to a gradual decrease in its diameter and to an increase in its luminosity. Other things being equal, this change in solar radiation must correspond to the very cold climate of pre-Cambrian and Paleozoic times, which is contrary to the data of geological investigations. To explain this contradiction an assumption was made that the atmospheric chemical composition was quite different in the past, increasing the greenhouse effect of the atmosphere. If this assumption is correct, it can be concluded that a random coincidence of variations in external factors (solar radiation and the chemical composition of the atmosphere), whose influence on the thermal regime is opposite and which compensated for each other, made it possible to preserve life on our planet for so long a time.

The probability of the biosphere's destruction in 1 to 2 billion years (i.e., long before the expiration of the solar radiation) as a result of the overheating of the earth is another conclusion from a concept concerning the solar radiation growth that was introduced by several authors.

The exchange of matter between deep layers of the lithosphere and the ocean–atmosphere system has exerted a great influence on the biosphere. It is thought that the ocean and atmosphere emerged as a result of the process of degassing of the upper mantle, in the course of which water vapor, CO_2, and other substances entered the surface layer of the earth. This process is a consequence of the heating of the lithosphere caused by the decay of a number of long-lived isotopes of radioactive elements. Undoubtedly, the rate of the degassing process was not constant. This is indicated by fluctuations in volcanic activity, which have varied considerably during geological history. Since volcanos eject a significant amount of CO_2 and other products, their activity exerted a great influence on the biosphere.

It follows from Chapter 2 that, during the Phanerozoic, the amount of atmospheric CO_2 varied by about a factor of ten, from several tenths to several hundredths of a percent of the atmospheric volume and the amount of oxygen by 5 times, in the range of 30–150% of its present mass.

Undoubtedly, variations in atmospheric chemical composition exerted a great influence on animate nature.

Of some interest is the study of the possibility of changing the atmospheric chemical composition to the limits that exclude the further existence of organisms. As is known, autotrophic plants, which are the main source of energy for the modern biosphere, usually can function in the range of CO_2 concentration from 0.01–0.02% to several percent.

For most animals there is an upper limit of CO_2 concentration for their environment; above this limit they die. In many cases this limit is several percent of the atmospheric volume.

Although various plants and animals can exist only in definite ranges of partial pressure of O_2 (which are very different, particularly for aerobic and anaerobic organisms), the question of fluctuations in atmospheric O_2 mass is probably of lesser significance for the stability of the biosphere than that of variations in CO_2 concentration. This can be attributed to the fact that the atmospheric O_2 mass far exceeds the CO_2 mass, the source of generation of free O_2 (the photosynthesis of autotrophic plants) being far less changeable than that of atmospheric CO_2. Therefore one can suppose that in the Phanerozoic the amount of CO_2 varied much more than the O_2 mass. The data from Chapter 2 confirm this assumption, although the actual differences between relative variations in CO_2 and O_2 masses must have been far greater than the values obtained, since those values are averages from the geological periods or epochs and are not characteristic of variations in the atmospheric chemical composition for shorter time intervals. For these reasons considerable variations in the atmospheric O_2 mass for relatively short intervals are highly improbable, whereas for the CO_2 mass, such a possibility undoubtedly existed. The evidence of this is a rapid growth of CO_2 concentration in the modern epoch due to the impact of man's economic activities.

As mentioned above, atmospheric CO_2 greatly influences the thermal regime by increasing the greenhouse effect. A decrease in CO_2 concentration below the modern one could result in temperature lowering that, as some computations show, for sufficiently low concentrations could lead to complete glaciation of the earth. Thus CO_2 variations exert both a direct influence on living organisms and an indirect one connected with climatic change.

Before the above-mentioned ecological catastrophes associated with the sun's evolution occur, the atmospheric chemical composition could vary, going beyond the limits of the possibility of the existence of living organisms. Carbon dioxide concentration variations in particular present danger in this respect.

As indicated above, the range of CO_2 concentrations within which the existence of most of the autotrophic plants is possible is relatively small: it allows the CO_2 mass to increase or decrease by approximately an order of magnitude relative to its mean value for the Phanerozoic. At present the CO_2 mass is noticeably lower than its mean value in the past and is apparently close to the lower limit of the interval indicated.

As seen from the data of Chapter 2, the process of decreasing the amount of CO_2 in the atmosphere started in the Cretaceous period and accelerated at the end of the Tertiary period. The probable reason for the decreasing CO_2 mass is the attenuation of volcanic activities, possibly caused by the exhaustion of the reserves of radioactive elements that produce heating of the lithosphere. From the available approximate estimates it follows that, if this process continues for about a million years (a period short from the viewpoint of the earth's history) one of the two ecological catastrophes will occur: the complete glaciation of the earth or the disappearance of autotrophic plants.

In this connection one should remember the opinion of Ronov (1976) that life on the earth can exist only with a sufficient level of decay of radioactive elements in the lithosphere.

It is assumed that in the Pleistocene the earth's biosphere was not far from destruction in the epochs of the greatest development of the last Quaternary glaciations, which advanced close to the critical latitude (the limit beyond which ice loses its stability and shifts toward the equator as in a self-propelled process).

Assuming the possibility of the disappearance of the biosphere in the not very distant future, one should return to the question of how the biosphere could be maintained for such a long period in the past. It is believed that the maintenance on the earth of a mean temperature within the narrow zone necessary for life for billions of years seems to be a random event, the probability of which is very low. To a considerable extent, the comparatively small changeability of the atmospheric chemical composition, whose variations could easily destroy all organisms or most of them, was also random.

As mentioned in some investigations, along with noticeable variations in mean volcanic activity over long intervals of time, short term variations far greater in amplitude occurred. The data on present-day CO_2 concentration growth indicate that there are no strong negative feedbacks in nature that could essentially decrease the growth of atmospheric CO_2 mass when its income is increased. In studying present changes in atmospheric chemical composition, one might note that now the anthropogenic production of CO_2 is much more intensive than the volcanic and other natural sources of CO_2. But the modern level of volcanic activity is far from the highest possible.

During the Phanerozoic, aerobic organisms seemed to be in permanent danger of being poisoned by CO_2 and other products of volcanic explosions, among which is the very toxic substance, carbon monoxide.

The decrease in atmospheric transparency accompanying a considerable increase in volcanic activity was another danger for living organisms. As the estimates made show, variations in volcanic activity over the last centuries led to a change in mean surface air temperature of several tenths of a degree.

Undoubtedly, during hundreds of millions of years far greater temperature variations must have been observed due to the coincidence in individual time intervals of many volcanic explosions. Rough estimates of this effect have led to the conclusion that, over individual decades, the mean air temperature could have been lowered by $5-10°$ as a result of this cause. Such a temperature reduction (or its sequel) appears to be a sufficient cause for the extinction of many species of stenothermal organisms (i.e., organisms that could not adapt to considerable temperature fluctuations in the environment). In this connection it has been hypothesized that increasing volcanic activity could have caused the changes of successive faunas in critical epochs of geological history (Budyko, 1971).

The question whether an increase in volcanic activity could result in greater temperature lowering, sufficient for most or all living organisms to be destroyed, is worth studying.

For complete elucidation of the problem it is necessary that a quantitative assessment of the probability of the changes in external factors that would exert a significant influence on the biosphere be carried out. Although obtaining these estimates is a matter for the future, the available restricted data enable one to suppose that, over the very long period of the biosphere's existence, this probability seemed to be high.

The uniqueness of the earth's biosphere

It follows from the above considerations that life in the universe is a rare phenomenon. This refers particularly to the highest life forms, whose development could have resulted in the creation of extraterrestrial civilizations.

Variations in natural conditions that accelerated the evolution process seem to be of great importance in the emergence of the highest forms of organisms, which are responsible for the creation of civilizations. The importance of this factor for the change of successive faunas is acknowledged by many investigators, among them Simpson (1944), who wrote that among the reptiles, birds, and mammals the percentage of the orders that emerged during periods of intensive elevations and orogenesis is too high to be attributed to random cases. It is possible that, on a hypothetical planet with conditions favorable for life that were constant for billions of years, the slowness of evolution could exclude the possibility of the emergence of sufficiently complicated organisms and, consequently, the creation of a civilization.

Thus the emergence of highly developed organisms apparently requires a definite changeability of external conditions, this changeability being within the limits of the narrow life zone of the organisms indicated. Such a condition lowers the probability of the emergence of extraterrestrial civilizations compared to that of extraterrestrial life.

The method widely used in assessing the number of planets in the universe where civilizations might exist is to multiply the number of stars by several coefficients, each less than unity, among which are those characterizing the probabilities of (a) the existence of planets near a star, (b) the presence on these planets of conditions suitable for life, (c) the emergence of life on these planets, (d) the progress of organisms up to the level of the emergence of highly developed civilizations, and (e) the long-term existence of these civilizations. The authors of similar assessments acknowledge the great difficulty of obtaining any valid estimates of these coefficients (with the partial exclusion of the first of them), but, with the enormous number of stars in the universe, they can, by using very small (in their opinion) values for every coefficient, obtain a rather large number for the civilizations existing now. For example, Sagan (1975) assumes that in our galaxy approximately one million planets have highly developed civilizations.

Agreeing with Simpson that such a method of computation is to a large measure arbitrary, we should indicate that the authors of these calculations have no idea of the presence of additional factors that considerably lower the probability of the existence of life (and, particularly, intelligent life) in the universe.

From studies of the stability of the earth's biosphere it follows that this stability is not very high, and the probability of the existence of life on a planet similar to the earth for billions of years is extremely low. Comparatively small variations in solar radiation and in volcanic activity seem to be sufficient for the destruction of all multicellular organisms or all the life on the earth.

One can cite other external factors that could exert a fatal influence on organisms. For example, the fall of large meteorites, the tracks of which have been worn on the earth's surface but are clearly defined on the surfaces of the moon, Mars, and Mercury, could produce variations in the state of the biosphere, approaching the limits of the life zone.

Without regarding this and other similar possibilities in more detail, we note that, from our point of view, the independent evolution of the sun and the weak dependence of the evolution of the deep layers of the lithosphere on the state of the biosphere are of most importance for the problem under consideration.

These factors considerably decrease the probability of the existence of the biosphere for a time sufficient for the emergence of a highly developed civilizations.

Therefore, it might be thought that the probability of life in the universe, particularly in its highest forms that are associated with the appearance of civilizations, is insignificant. This conclusion is confirmed by the negative results of life detection experiments on the moon and Mars, the failure of attempts to pick up signals from extraterrestrial civilizations, and the absence of reliable information on past contacts between mankind and the inhabitants of other planets.

This conclusion is based on studies of the earth's biosphere, the results of which have been stated in previous works by the author (Budyko, 1971, 1974, 1977a). A similar conclusion has recently been drawn by Hart (1978, 1979), who evaluated the possibility of the maintenance of conditions favorable for life on the planets revolving around different stars.

In the first of these articles Hart has constructed a numerical model of the evolution of the chemical composition of the earth's atmosphere whose application shows that the long maintenance of conditions favorable for the existence of organisms is possible only for a specific range of distance between the earth and the sun. An increase in this distance of the order of 1% could have led to complete glaciation of the planet in the remote past. A slight decrease in this distance could lead to an increase of the planet's surface temperature to the boiling point of water. Then the disappearance of the hydrosphere, where the CO_2 emitted from the earth's crust is being absorbed, would result in a rapid further temperature rise, as occurred on Venus.

Hart's second article treats the question of the existence of "zones of life" (i.e., places where the existence of living organisms is possible on planets similar to the earth) for stars of different types. Using the above model, Hart has found that zones of life could not exist for long near many stars. For the stars that could have long-term zones of life, the volume of these zones, which is determined by the closest and farthest approach of the planet to the star, is very small in every case.

Although Hart's model of the evolution of the atmosphere is rather schematic (see Chapter 2), it correctly simulates many features of this process. In addition, Hart has shown that the conclusions on the properties of the life zone depend comparatively little on the possible errors in the parameters introduced into his model. For this reason, Hart's results can be considered one of the evidences that life in the universe is a very rare phenomenon.

In view of these considerations, the earth's biosphere is unique in two respects. First, the very fact of its long existence under conditions that repeatedly could have led to the destruction of living things on our planet is extraordinary. Second, the capability of certain species of living organisms, namely mankind, of prolonging the existence of the biosphere by averting the processes that threaten its further life, is also exceptional.

The future of the biosphere

The first result of the global impact produced by man on the biosphere that is connected with restoring the ancient CO_2-rich atmosphere is an inadvertent consequence of economic activities. This change in the composition of the atmosphere has increased the stability of the biosphere and decreased the possibility of its complete or partial destruction by glaciations.

If it is assumed that the concentration of atmospheric CO_2 will decrease again after all the available resources of fossil fuel are used, one might think that at that time the level of energy production from other sources will be high enough to maintain the optimum air temperature near the earth's surface.

In recent years new technological methods have come into view that can modify the global climate, either by increasing the mean temperature in the lower air layers (introducing water vapor, chlorofluoromethane, or other gases that absorb longwave radiation in the stratosphere) or by decreasing this temperature (adding to the aerosol layer in the stratosphere).

Undoubtedly, in the future it will be possible to influence the climate to a much greater extent and to control it for the good of all mankind. With such technological means we can greatly increase the stability of the biosphere and considerably lengthen the time of its existence.

6.4 CONCLUSION

The question of the climatic conditions of the future is part of the general problem of human impact on the environment, which is now attracting great attention. Of this problem, Fedoseev (1978) writes: "For the first time now humanity at large has come across a problem of averting actions and processes which could drastically disturb the normal state of the biosphere and thereby undermine natural prerequisites of social being."

Mentioning a well known statement by Karl Marx, that culture leaves behind itself a desert if it develops spontaneously and is not directed deliberately, Fedorov says

The character of interaction of human society and nature attracts an increasing interest in the last few decades and at the same time it alarms scientists and laymen alike in all corners of the world. Numerous articles and speeches relating to this subject often state an opinion that a developed society ruins nature, destroys the environment in which it exists and from which it derives all the necessary resources. Nature is being destroyed by the very fact of society's rapid development, its technology and mainly by unconsidered actions. And thereby the society deprives itself of prospects and possibilities for subsequent existence.

Fedorov remarks that to eliminate the danger of a global ecological crisis "mankind must form nature on the global scale." And this can be achieved

only by an organized society which is able to control its actions on the whole and its interactions with nature on the global scale. . . . The matter is whether humanity would be able in the nearest decades to control and direct its own development. We think that this very problem defines a critical situation of the next stage of mankind's progress (Fedorov, 1977).

Taking account of these considerations, one must acknowledge that the most important task of modern ecological research is the study, aimed at averting adverse effects, of the consequences of human economic activities relating to the environment.

This book has shown that man's economic activities, as they are connected with the ever increasing burning of fossil fuel and the cutting of forests, lead to changes in the atmospheric chemical composition and the changes rapidly become greater. An increase in atmospheric CO_2 has already resulted in a mean air temperature rise near the earth's surface of a few tenths of a degree. With further progress in economic activity the mean air temperature will rise by 2–3° in several decades and later it will increase even more. This will lead to the destruction of polar sea ice, which will make climatic changes in high and middle latitudes in the Northern Hemisphere more pronounced.

Calculations on the basis of climatic models and paleoclimatic information enable one to construct maps defining climatic conditions in the future. With the data of these maps it is possible to evaluate variations in different components of the natural conditions in response to an increase in CO_2 in the atmosphere. The estimates obtained indicate that these variations could be quite large over the entire globe, especially in high and middle latitudes.

It is essential for understanding the present changes in the state of the biosphere that, with an increase in atmospheric CO_2 concentration, the atmosphere returns to the composition typical of the Tertiary period, when climatic conditions were far warmer and plant productivity was higher than at present.

Since the atmosphere lost CO_2 for the last hundred million years, directly threatening the existence of the biosphere due to a decrease in the productivity of autotrophic plants and the possible complete glaciation of the earth, the present anthropogenic impact on the biosphere seems to be a favorable factor that eliminates the indicated threat.

However, it must be mentioned that although many aspects of global warming could be favorable for mankind (a rise in the productivity of autotrophic plants, better usage of land in cold climate zones, etc.), a number of difficulties can also inevitably arise.

The major difficulty lies in the necessity of adjusting, in a relatively brief period of time, many branches of the economy to the conditions of a rapidly changing climate and other aspects of natural conditions.

It is quite evident that there is now a need for a detailed study of the major consequences of global climatic change so that they can be considered while planning economic activities.

In connection with this, it is necessary to organize, as quickly as possible, broad interdisciplinary investigations of anthropogenic climatic changes and their impact on natural conditions. In view of the global charactor of the processes involved, the proposed studies must be extensive and based on international co-operation.

References*

Adhèmar, J. F. (1842). "Les Révolutions de la Mer—Déluges Périodiques." Paris.
Ahlman, H. W. (1953). Glacier variations and climatic fluctuations, *Amer. Geogr. Soc. Ser. 3*, 51.
Albrecht, F. (1947). Die Aktiongebiete des Wasser- und Wärmehaushaltes der Erd-oberfläche, *Z. Meteorol.* **1**(4–5), 97–109.
Allen, C. W. (1958). Solar radiation, *Q. J. R. Meteorol. Soc.* **84**(362), 307–318.
Alt, F. (1929). Der Stand des meteorologischen Strahlungsproblems, *Meteorol. Z.* **46**(12).
Alyea, F. N. (1972). Numerical simulation of an ice age paleoclimate, Atmospheric Science Paper No. 193. Colorado State Univ., Fort Collins, Colorado.
Andronova, N. G., Karol', I. L., and Frol'kis, V. A. (1977). The method of small disturbances in a stationary semi-empirical model of atmospheric mean zonal heat balance, *Proc. State Hydrological Inst.* No. 247, 55–64. (R)
Angell, J. K., and Korshover, J. (1977). Estimate of the global change in temperature, surface to 100 mb, between 1958 and 1975, *Mon. Weather Rev.* **105**(4), 375–385.
Angell, J. K., and Korshover, J. (1978). Global temperature variation, surface to 100 mb: an update into 1977, *Mon. Weather Rev.* **106**(6), 755–770.
Ångström, A. (1962). Atmospheric turbidity, global illumination and planetary albedo of the earth, *Tellus* **14**(4), 435–450.
Ångström, A. (1965). The solar constant and the temperature of the Earth, *in* "Progress in Oceanography," Vol. 3, pp. 1–5. Pergamon, Oxford.

* (R) indicates Russian language references

288

Ångström, A. (1969). Apparent solar constant variations and their relation to the variability of atmospheric transmission, *Tellus* **21**(2), 205–218.

Arrhenius, S. (1896). On the influence of the carbonic acid in the air upon the temperature of the ground, *Philos. Mag.* **41**, 237–275.

Arrhenius, S. (1903). "Lehrbuch der kosmischen Physik," Vol. 2. Hirzel, Leipzig.

Atlas (1964). "Physico-geographical Atlas of the World." GUK, Moscow. (R)

Atlas (1974). "Atlas of the Oceans." Vol. 1. VMF, Moscow. (R)

Atwater, M. A. (1970). Planetary albedo changes due to aerosols, *Science* **170**(3953), 64–66.

Atwater, M. A. (1977). Urbanization and pollutant effects on the thermal structure in four climatic regimes, *J. Appl. Meteorol.* **16**, 888–895.

Augustsson, T., and Ramanathan, V. (1977). A radiative-convective model study of the CO_2 climate problem, *J. Atmos. Sci.* **34**, 448–451.

Baes, C. F. *et al.* (1976). "The Global Carbon Dioxide Problem." Oak Ridge National Laboratory.

Baldwin, B. *et al.* (1976). Stratospheric aerosols and climatic change, *Nature (London)* **263**, 551–555.

Barnett, T. P. (1978). Estimating variability of surface air temperature in the Northern Hemisphere, *Mon. Weather Rev.* **106**(9), 1353–1367.

Barrett, E. (1971). Depletion of short-wave irradiance at the ground by particles suspended in the atmosphere, *Sol. Energy* **13**(3), 323–337.

Baur, F., and Philipps, H. (1934). Der Wärmehaushalt der Lufthülle der Nordhalbkugel, *Gerlands Beitr. Geophys.* **43**, 160–207.

Berger, A. L. (1973). "Théorie Astronomique des Paléoclimats," Vols. 1 and 2. Université Catholique de Louvain, Louvain, Belgium.

Berger, A. L. (1975). La paléoclimatologie quantitative du Quaternaire—Part I, Part II, *Rev. Quest. Sci.* **146**(2), 167–179; **146**(3), 273–293.

Berger, A. L. (1976). Obliquity and precession for the last 5,000,000 years, *Astron. Astrophys.* **51**, 127–135.

Berger, A. L. (1977). Power and limitation of an energy-balance climate model as applied to the astronomical theory of palaeoclimates, *Palaeogeogr. Palaeoclimatol. Palaeoecol.* **21**, 227–235.

Berger, A. L. (1978). La théorie astronomique des paléoclimats, une nouvelle approche, *Bull. Soc. Belge Géol.* **87**, 9–25.

Berkner, L. V., and Marshall, L. C. (1965a). On the origin and rise of the oxygen concentration in the Earth's atmosphere, *J. Atmos. Sci.* **22**(3), 225–261.

Berkner, L. V., and Marshall, L. C. (1965b). Oxygen and evolution, *New Sci.* **28**(469), 415–419.

Berkner, L. V., and Marshall, L. C. (1966). Limitation on oxygen concentration in a primitive planetary atmosphere, *J. Atmos. Sci.* **23**(2), 133–143.

Berlyand, M. E., and Kondratév, K. Ya. (1972). "Cities and Climate of the Planet." Gidrometeoizdat, Leningrad. (R)

Berlyand, T. G., and Strokina, L. A. (1975). Cloudiness regime over the terrestrial globe, *Proc. Main Geophys. Observ.*, No. 338, 3–20. (R)

Blinova, E. N. (1947). On the mean annual temperature distribution in the earth's atmosphere with consideration of continents and oceans, *Izv. Akad. Nauk USSR Ser. Geogr. Geofiz.* **11**(1), 3–13. (R)

Bogorov, V. G. (1969). "The Life of the Ocean." Znanie, Moscow. (R)

Bolin, B. (1975). "Energy and Climate." Stockholm.

Bolin, B. (1977). Changes in land biota and their importance for the carbon cycle, *Science* **196**, 613–619.

Bolin, B., and Keeling, C. D. (1963). Large-scale atmospheric mixing as deduced from the seasonal and meridional variations of carbon dioxide, *J. Geophys. Res.* **68**(13), 3899–3920.

Borzenkova, I. I. (1974). On the possible influence of volcanic dust on the radiation and thermal regime, *Proc. Main Geophys. Observ.* No. (307), 36–42.   (R)

Borzenkova, I. I. *et al.* (1976). Changes of air temperature of the Northern Hemisphere. *Meteorol. Gidrol.* (7), 27–35.   (R)

Bossolasco, M. *et al.* (1964). Globalstrahlung und Sonnenaktivität, *Pure Appl. Geophys.* **57**, 221–224.

Bowen, H. J. M. (1966). "Trace Elements in Biochemistry." Academic Press, New York.

Brinkman, R. T. (1969). Dissociation of water vapour and evolution of oxygen in the terrestrial atmosphere, *J. Geophys. Res.* **74**(23), 5355–5368.

Broecker, W. S. (1975). Climatic change: Are we on the brink of a pronounced global warming? *Science* **189**, 460–463.

Brooks, C. E. P. (1949). "Climate through the Ages," 2nd ed. Ernst Benn, London.

Brückner, E. (1890). Klimaschwankungen seit 1700, nebst Bemerkungen über die Klimaschwankungen der Deluvialzeit, *Geogr. Abhandl. von Peck Vienna* **4**(2).

Bryson, R. A. (1968). All other factors being constant . . . , *Weatherwise* **21**(2), 56–62.

Buchardt, B. (1978). Oxygen isotope palaeotemperatures from the Tertiary period in the North Sea area, *Nature (London)* **275**, 121–123

Budyko, M. I. (1946). Methods of determining natural evaporation, *Meteorol. Gidrol.* No. 3, 3–15.   (R)

Budyko, M. I. (1947). On the water and heat balances of the land surface. *Meteorol. Gidrol.* No. 5, 49–64.   (R)

Budyko, M. I. (ed.). (1955). "Atlas of the Heat Balance." Izd. GGO, Leningrad.   (R)

Budyko, M. I. (1956). "Heat Balance of the Earth's Surface." Gidrometeoizdat, Leningrad   (R); (1958). transl. by N. A. Stepanova, MGA 13E-286, U.S. Weather Bureau, Washington, D.C., 11B–25.

Budyko, M. I. (1961). On the thermal zonality of the earth, *Meteorol. Gidrol.* **11**, 7–14.   (R)

Budyko, M. I. (1962a). Certain methods of influencing the climate, *Meteorol. Gidrol.* **2**, 3–8.   (R)

Budyko, M. I. (1962b). Polar ice and climate, *Izv. Akad. Nauk USSR Ser. Geogr.* **6**, 3–10.   (R)

Budyko, M. I. (ed.) (1963). "Atlas of the Heat Balance of the Earth" Mezhduvedomstvennyi geofizicheskii komitet, Moscow.   (R)

Budyko, M. I. (1964). On the theory of photosynthesis effects of climate factors, *Rep. Akad. Nauk USSR* **158**(2), 331–337.   (R)

Budyko, M. I. (1968a). On the origin of ice ages, *Meteorol. Gidrol.* No. 11, 3–12.   (R)

Budyko, M. I. (1968b). Radiation factors of present-day climatic change, *Izv. Akad. Nauk USSR Ser. Geogr.* No. 5, 36–42.   (R)

Budyko, M. I. (1969a). "Changes of Climate." Gidrometeoizdat, Leningrad.   (R)

Budyko, M. I. (1969b). "Polar Ice and Climate." Gidrometeoizdat, Leningrad.   (R)

Budyko, M. I. (1970). Comments on: A global climatic model based on the energy balance of the Earth–atmosphere system, *J. Appl. Meteorol.* **9**(2), 310–311.

Budyko, M. I. (1971). "Climate and Life." Gidrometeoizdat, Leningrad   (R); (1974). English edition edited by David H. Miller. Academic Press, New York.

Budyko, M. I. (1972a). Comments on: Intransitive model of the Earth–atmosphere–ocean system, *J. Appl. Meteorol.* **11**(7), 1150–1151.

Budyko, M. I. (1972b). "Man's Influence on Climate." Gidrometeoizdat, Leningrad.   (R)

Budyko, M. I. (1974). "Climatic Changes." Gidrometeoizdat, Leningrad (R); (1977). English Translation by American Geophysical Union, Washington, D.C.

Budyko, M. I. (1975). The dependence of mean air temperature on solar radiation variations, *Meteorol. Gidrol.* **10**, 3–10. (R)

Budyko, M. I. (1977a). "Global Ecology." Mysl', Moscow. (R)

Budyko, M. I. (1977b). Studies of present-day climatic changes, *Meteorol. Gidrol.* **11**, 42–57. (R)

Budyko, M. I. (1979a). "The Problem of Carbon Dioxide." Gidrometeoizdat, Leningrad. (R)

Budyko, M. I. (1979b). Semiempirical model of atmospheric thermal regime and real climate, *Meteorol. Gidrol.* **4**, 5–17. (R)

Budyko, M. I., and Drozdov, O. A. (1953). Characteristics of the moisture circulation in the atmosphere, *Izv. Akad. Nauk USSR Ser. Geogr.* (4), 5–14. (R)

Budyko, M. I., and Drozdov, O. A. (1976). On the causes of changes in water turnover, *Water Resources* (6), p. 35–44. (R)

Budyko, M. I., and Gandin, L. S. (1964). Taking account of laws of atmospheric physics in agrometeorological studies, *Meterol. Gidrol.* **11**, 3–11. (R)

Budyko, M. I., and Gandin, L. S. (1965). On the theory of photosynthesis in vegetation cover layer, *Rep. Akad. Nauk USSR* **164**(2), 454–457. (R)

Budyko, M. I., and Gandin, L. S. (1966). The influence of climatic factors on vegetation cover, *Izv. Akad. Nauk USSR Ser. Geogr.* **1**, 3–10. (R)

Budyko, M. I., and Pivovarova, Z. I. (1967). The influence of volcanic eruptions on solar radiation incoming to the Earth's surface, *Meteorol. Gidrol.* **10**, 3–7. (R)

Budyko, M. I., and Ronov, A. B. (1979). Atmospheric evolution in the Phanerozoic, *Geochemistry* **5**, 643–653. (R)

Budyko, M. I., and Vasishcheva, M. A. (1971). Effect of astronomical factors on Quaternary glaciations, *Meteorol. Gidrol.* **6**, 37–47. (R)

Budyko, M. I., and Vinnikov, K. Ya. (1973). Contemporary climatic changes, *Meteorol. Gidrol.* **9**, 3–13. (R)

Budyko, M. I., and Vinnikov, K. Ya. (1976). Global warming, *Meteorol. Gidrol.* **7**, 16–26. (R)

Budyko, M. I., Drozdov, O. A., L'vovich, M. I., Sapozhnikova, S. A., and Yudin, M. I. (1952). "Changing the Climate in Connection with the Plan for Transforming Nature in the Drought Regions of the USSR." Gidrometeoizdat, Leningrad. (R)

Budyko, M. I., Drozdov, O. A., and Yudin, M. I. (1966). Man's economic activity impact on climate, *in* "Present-day Climatology Problems," pp. 435–448. Gidrometeoizdat, Leningrad. (R)

Budyko, M. I. *et al.* (1978a). "Heat Balance of the Earth." Gidrometeoizdat, Leningrad. (R)

Budyko, M. I. *et al.* (1978b). Foregoing climatic changes, *Izv. Akad. Nauk USSR Ser. Geogr.* **6**, 5–20. (R)

Burdecki, F. (1964). Phenomena after volcanic eruptions, *Weather* **19**(4), 113–114.

Burtsev, A. I. (1955). Water exchange elements on the European USSR territory, *Proc. CIP* **38**(55). (R)

Butzer, K. W. (1964). "Environment and Archeology." Methuen, London

Byutner, E. K. (1961). On the time for establishing a stable amount of oxygen in the atmospheres of the planets containing water vapour, *Rep. Akad. Nauk USSR* **138**, 1050–1053. (R)

Callendar, G. S. (1938). The artificial production of carbon dioxide and its influence on temperature, *Q. J. R. Meteorol. Soc.* **64**(27), 223–240.

Cess, R. D. (1976). Climatic change: An appraisal of atmospheric feedback mechanisms employing zonal climatology, *J. Atmos. Sci.* **33**(10), 1831–1843.

Cess, R. D. (1978). Biosphere–albedo feedback and climate modeling, *J. Atmos. Sci.* **35**(9), 1765–1768.

Cess, R. D., and Wronka, J. C. (1979). Ice ages and the Milankovich theory: A study of interactive climate feedback mechanisms, *Tellus* **39**(3), 185–192.

Chamberlin, T. C. (1897). A group of hypotheses bearing on climatic changes, *J. Geol.* **5**, 653–683.

Chamberlin, T. C. (1898). The influence of great epochs of limestone formation upon the constitution of the atmosphere, *J. Geol.* **6**, 609–621.

Chamberlin, T. C. (1899). An attempt to frame a working hypothesis of the cause of glacial periods on an atmospheric basis, *J. Geol.* **7**, 545–584.

Charlson, R. J., and Pilat, M. J. (1969). Climate: The influence of aerosols, *J. Appl. Meteorol.* **8**(6), 1001–1002.

Chizhov, O. P., and Tareeva, A. M. (1969). Possibilities of estimating the ice regime of the Arctic basin and its variations, *Mater. Glaciolog. Stud.* **15**, 57–73.   (R)

Chylek, P., and Coakley, J. A. Jr. (1975). Analytical analysis of a Budyko-type climate model, *J. Atmos. Sci.* **32**(4), 676–679.

CIAP (1975). The Natural and Radiatively Perturbed Troposphere, CIAP Monograph 4, Department of Transportation, Washington, D.C.

CLIMAP Project Members (1976). The surface of the ice-age earth, *Science* **191**, 1131–1138.

Cloud, P. E. (1974). Atmosphere, development of, "Encyclopaedia Britannica," 15th ed., pp. 313–319.

Coakley, J. A. (1979). A study of climate sensitivity using a simple energy balance model, *J. Atmos. Sci.* **36**(2), 260–269.

Damon, P. E., and Kunen, S. M. (1976). Global cooling? *Science* **193**(4252), 447–453.

Davitaya, F. F. (1965). On the possibility of the influence of atmospheric dust on shrinkage of glaciers and climate warming, *Izv. Akad. Nauk USSR Ser. Geogr.* No. 2, 3–23.   (R)

Department of Transportation (1975). Effect of stratospheric pollution by aircraft. Report of Findings. Department of Transportation, Washington, D.C.

Dines, W. H. (1917). The heat balance of the atmosphere, *Q. J. R. Meteorol. Soc.* **43**(151).

Donn, W. L., and Shaw, D. M. (1966). The heat budgets of an ice-free and an ice-covered Arctic Ocean, *J. Geophys. Res.* **71**(4), 1087–1093.

Dorodnitsyn, A.A., Izvekov, B. I., and Shvets, M. E. (1939). Mathematical theory of general circulation, *Meteorol. Gidrol.* (4), 32–42.   (R)

Dorst, J. (1965). "Avant que nature meure." Switzerland.

Drozdov, O. A., and Grigor'eva, A. S. (1963). "Water Exchange in the Atmosphere." Gidrometeoizdat, Leningrad.   (R)

Drozdov, O. A., and Grigor'eva, A. S. (1971). "Long-term Cyclic Fluctuations of Atmospheric Precipitation over the USSR Territory." Gidrometeoizdat, Leningrad.   (R)

Drozdov, O. A. *et al.* (1974). "World Water Balance and Water Resources of the Earth." Gidrometeoizdet, Leningrad.   (R)

Dwyer, H. A., and Peterson, T. (1973). Time-dependent global energy modeling, *J. Appl. Meteorol.* **12**(1), 36–42.

Dyer, A. J. (1974). The effect of volcanic eruptions on global turbidity and an attempt to detect long-term trends due to man, *Q. J. R. Meteorol. Soc.* **100**, 563–571.

Dyer, A. J., and Hicks, B. B. (1965). Stratospheric transport of volcanic dust inferred from solar radiation measurements, *Nature (London)* **208**(5006), 131–133.

Dyer, A. J., and Hicks, B. B. (1968). Global spread of volcanic dust from the Bali eruption of 1963, *Q. J. R. Meteorol. Soc.* **94**(402), 545–554.

Efimova, N. A. (1977). "Radiative Factors of Vegetation Cover Productivity." Gidrometeoizdat, Leningrad. (R)

Ellis, J., and vonder Haar, T. H. (1976). Zonal average earth radiation budget measurements from satellites for climate studies. Atmospheric Science Paper 240, Colorado State Univ., Fort Collins, Colorado.

Ellsaesser, H. W. (1975). The upward trend in airborne particulates that isn't, in "The Changing Global Environment," pp. 235–272, Reidel, Dordrecht, Holland.

Emiliani, C. (1955). Pleistocene temperatures, J. Geol. 63(6), 538–578.

Emiliani, C. (1966). Isotopic paleotemperatures, Science 154, 851–857.

Ensor, D. S. et al. (1972). Influence of atmospheric aerosol on albedo, J. Appl. Meteorol. 10(6), 1303–1306.

Ewing, D. M., and Donn, W. L. (1956). A theory of ice ages, I, Science 123, 1061–1066.

Ewing, D. M., and Donn, W. L. (1958). A theory of ice ages, II, Science 127, 1159–1162.

Faegre, A. (1972). An intransitive model of the Earth–atmosphere–ocean system, J. Appl. Meteorol. 11(1), 4–6.

Fairbridge, R. W. (1967). Ice-age theory, in "The Encyclopaedia of Atmospheric Sciences and Astrogeology," pp. 462–474. Van Nostrand-Reinhold, Princeton, New Jersey.

Fedorov, E. K. (1977). "Ecological Crisis and Social Progress." Gidrometeoizdat, Leningrad. (R)

Fedorov, E. K. (1979). Climate change and the strategy of mankind, Meteorol. Gidrol. 7, 12–24. (R)

Fedoseev, N. P. (1978). "Dialectic of Contemporary Epoch," 3rd ed. Nauka, Moscow. (R)

Fletcher, J. O. (1966). The Arctic heat budget and atmospheric circulation, Proc. Symp. Arctic Heat Budget Atmos. Circulat. pp. 23–43. Rand Corp., Santa Monica, California.

Flint, R. F. (1957). "Glacial and Pleistocene Geology." Wiley, New York.

Flohn, H. H. (1963). Zur meteorologischen Interpretation der pleistozänen Klimaschwänkungen, Eiszeitalter Ggw. 14, 153–160.

Flohn, H. H. (1964). Grundfragen der Paläoklimatologie im Lichte einer theoretischen klimatologie, Geol. Rundsch. 54(5), 504–515.

Flohn, H. H. (1969). Ein geophysikalisches Eiszeit-Modelle, Eiszeitalter Ggw. 20, 204–231.

Flohn, H. H. (1977a). Climate and energy. A scenario to the 21st century problem, Clim. Change 1(1), 5–20.

Flohn, H. H. (1977b). Man-induced changes in heat budget and possible effects on climate, in "Global Chemical Cycles and Their Alterations by Man," Dahlem Konferenzen, pp. 207–224. Berlin.

Flohn, H. H. (1978). Estimates of a combined greenhouse effect as a background for a climate scenario during global warming, IIASA workshop on carbon dioxide, climate, and society, in "Carbon Dioxide, Climate, and Society." pp. 227–238. Pergamon, Oxford.

Flowers, E. C., and Viebrock, H. J. (1965). Solar radiation: an anomalous decrease of direct solar radiation, Science 148(3669), 493–494.

Frederiksen, J. S. (1976). Nonlinear albedo–temperature coupling in climate models, J. Atmos. Sci. 33(12), 2267–2272.

Fuchs, V. E., and Patterson, T. T. (1947). The relation of volcanicity and orogeny to climate change, Geol. Mag. 84(6), 321–333.

Gaevskaya, G. N. (1972). Absorption of short-wave radiation by aerosol, Probl. Phis. Atmos. 10, 86–91. (R)

Gal-Chen, T., and Schneider, S. H. (1976). Energy balance climate modeling: Comparison of radiative and dynamic feedback mechanisms, Tellus 28, 108–121.

Gal'tsov, A. P. (1961). The meeting on the problem of changing climate, *Izv. Akad. Nauk. USSR Ser. Geogr.* No. 5, 128–133. (R)

Gal'tsov, A. P., and Chaplygina, A. S. (1962). The second meeting on the problem of changing climate. *Izv. Akad. Nauk USSR Ser. Geogr.* No. 5, 184–187. (R)

Gandin, L. S., Il'in, B. M., and Rukhovets, L. V. (1973). On the influence of variations of external parameters on the atmospheric thermal regime, *Proc. Main Geophys. Observ.* 315, 21–38. (R)

Garrels, R. M. (1975). "The Circulation of Carbon, Oxygen and Sulfur During Geological Time." Nauka, Moscow. (R)

Garrels, R. M., and Mackenzie, F. T. (1971). "Evolution of Sedimentary Rocks." Norton, New York.

Garrels, R. M. *et al.* (1975). "Chemical Cycles and the Global Environment." Kaufman, Los Altos, California.

Gates, W. L. (1975). The January global climate simulated by a two-level general circulation model: a comparison with observation, *J. Atmos. Sci.* 32(3), 449–477.

Gates, W. L. (1976a). Modeling the ice-age climate, *Science* 191, 1138–1144.

Gates, W. L. (1976b). The numerical simulation of ice-age climate with a global general circulation model, *J. Atmos. Sci.* 33(10), 1844–1873.

Gates, W. L. (1979). The physical basis of climate, *World Climate Conf.* pp. 71–84. WMO, Geneva.

Gerasimov, I. P. (1970). The nature and development of primitive society, *Izv. Akad. Nauk USSR Ser. Geogr.* 9, 5–8. (R)

Gerasimov, I. P. (1979). Climates in past geological epochs, *Meteorol. Gidrol.* 7, 37–53. (R)

Gerasimov, I. P., and Markov, K. K. (1939). "The Ice Age in the USSR Territory." Izd. Akad. Nauk USSR, Moscow. (R)

Golitsyn, G. S., and Mokhov, I. I. (1978a). On the stability and extremum properties of climate models, *Izv. Akad. Nauk USSR Phys. Atmos. Ocean* 14(4), 378–387. (R)

Golitsyn, G. S., and Mokhov, I. I. (1978b). Assessment of sensitivity and the role of cloudiness in simple climate models, *Izv. Akad. Nauk USSR Phys. Atmos. Ocean* 14(8), 803–813. (R)

Gol'tsberg, I. A. (1952). Expedition for the study of atmospheric turbulence in forest strips, *Proc. Main Geophys. Observ.* No. 29, 5–10. (R)

Gordon, H. B., and Davies, D. R. (1974). The effect of changes in solar radiation on climate, *Q. J. R. Meteorol. Soc.* 13(7), 752–759.

Grigor'ev, A. A. (1956). "The Sub-Arctic." Geografgiz, Moscow. (R)

Grigor'ev, A. A., and Budyko, M. I. (1959). Classification of the climates of the USSR, *Izv. Akad. Nauk USSR Ser. Geogr.* No. 3, 3–19. (R)

Grigor'eva, A. S., and Strokina, L. A. (1977). Variations in temperature course at high latitudes of the Northern Hemisphere, *Proc. State Hydrological Inst.* No. 247, 114–118. (R)

Grobecker, A. J. *et al.* (1974). The effects of stratospheric pollution by aircraft, CIAP Reports of findings. DOT-IST-75-50, Washington, D.C.

Grosswal'd, M. G., and Kotlyakov, V. M. (1978). Foregoing climate changes and the fate of glaciers, *Izv. Akad. Nauk USSR Ser. Geogr.* No. 6, 21–32. (R)

Hameed, S. *et al.* (1979). Sensitivity of the predicted $CO-OH-CH_4$ perturbation to tropospheric NO_x concentrations, *J. Geophys. Res.* 84, 763–768.

Hansen, J. E., Wang, W. Ch., and Lacis, A. (1978). Mount Agung eruption provides test of a global climatic perturbation, *Science* 199, 1065–1068.

Hantel, M. (1974). Polar boundary conditions in zonally averaged global climate models, *J. Appl. Meteorol.* **13**(7), 752–759.

Harley, W. S. (1978). Trends and variations of mean temperature in the lower troposphere, *Mon. Weather Rev.* **106**(3), 413–416.

Hart, M. H. (1978). The evolution of the atmosphere of the earth, *Icarus* **33**, 23–39.

Hart, M. H. (1979). Habitable zones about main sequence stars, *Icarus* **37**, 351–357.

Hays, J. D., Imbrie, J., and Schackleton, N. J. (1976). Variations in the Earth's orbit: Pacemaker of the ice-ages, *Science* **194**, 1121–1132.

Held, I. M., and Suarez, M. J. (1974). Simple albedo feedback models of the ice-caps, *Tellus* **26**, 613–629.

Holloway, J. L., and Manabe, S. (1971). A global general circulation model with hydrology and mountains, *Mon. Weather Rev.* **99**(5), 335–370.

Houghton, H. G. (1954). On the annual heat balance of the Northern Hemisphere, *J. Meteorol.* **11**(1), 1–9.

Humphreys, W. J. (1913). Volcanic dust and other factors in the prediction of climatic changes and their possible relation to ice age, *J. Franklin Inst.* **176**, 131.

Humphreys, W. J. (1929). "Physics of the Air," 2nd ed. McGraw-Hill, New York.

International Study Conference in Stockholm (1975). The physical basis of climate and climate modelling, Report of the *Int. Study Conf. Stockholm, July 29–August 10, 1974.* GARP Publ. Ser., No. 16.

Izrael, Yu. A. (1979). Climate monitoring and service for collection of climatic data necessary for determining climate change and variability, *Meteorol. Gidrol.* No. 7, 54–67. (R)

Jerison, H. J. (1973). "Evolution of the Brain and Intelligence." Academic Press, New York.

Junge, C. (1963). "Atmospheric Chemistry and Radioactivity," pp. 209–238. Academic Press, New York.

Kagan, R. L., and Vinnikov, K. Ya. (1970). On the predetermination of heat income in numerical experiments with the help of thermotropic model, *Proc. Main Geophys. Observ.* No. 256, 98–107. (R)

Kalitin, N. N. (1920). On the time of the onset of the optical anomaly in 1912, *Izv. GFO* **1**, 11–17. (R)

Karol', I. L. (1972). "Radioactive Isotopes and Global Transfer in the Atmosphere." Gidrometeoizdat, Leningrad. (R)

Karol', I. L. (1973). The size of radioactive aerosol and its transfer in the troposphere and stratosphere, *Meteorol. Gidrol.* **1**, 28–38. (R)

Karol', I. L. (1977). Changes in global content of stratospheric aerosol and their relation to variations in mean direct solar radiation and in temperature at the Earth's surface, *Meteorol. Gidrol.* **3**, 32–40. (R)

Karol', I. L., and Pivovarova, Z. I. (1978). Relationship between changes in stratospheric aerosol content and solar radiation fluctuations, *Meteorol. Gidrol.* **9**, 35–42. (R)

Keeling, C. D., and Bacastow, R. B. (1977). Impact of industrial gases on climate, *in* "Energy and Climate. Studies in Geophysics, pp. 72–95. National Academy of Sciences, Washington, D.C.

Kellogg, W. W. (1975). Climate change and the influence of man's activities on the global environment, *in* "The Changing Global Environment," pp. 13–23, Reidel, Dordrecht, Holland.

Kellogg, W. W. (1977). The influence of man's activities on climate, *Bull. WMO* **26**(4), 285–299. (R); (1978). **28**(1), 3–12. (R)

Kennet, J. P., and Thunell, R. C. (1975). Global synchronism and increased Quaternary explosive volcanism, *Science* **187**, 497–503.

Kester, D. R., and Pytkowicz, R. M. (1977). Natural and anthropogenic changes in the global carbon dioxide system, *in* "Global Chemical Cycles and Their Alterations by Man," pp. 99–120, Dahlem Konferenzen, Berlin.

Khromov, S. P. (1973). Solar cycles and climate, *Meteorol. Gidrol.* No. 9. 93–111   (R)

Kimball, H. H. (1918). Volcanic eruptions and solar radiation intensities, *Mon. Weather Rev.* **46**(8), 355–356.

Kochin, N. E. (1936). Construction of the model of zonal atmospheric circulation, *Proc. Main Geophys. Observ.* No. 10, 3–27.   (R).

Köppen, W., and Wegener, A. (1924). "Die Klimate der Geologischen Vorzeit." Berlin.

Kondrat'ev, K. Ya. *et al.* (1973). "The Influence of Aerosol on the Transfer of Radiation: Possible Climatic Consequences." Izd. LGU, Leningrad.   (R)

Kovda, V. A. (1969). The problems of biological economic productivity of land, *in* "General Theoretical Problems of Biological Productivity," pp. 8–23. Nauka, Leningrad.   (R)

Kukla, G. J. *et al.* (1977). New data on climatic trends, *Nature (London)* **270**, 573–580.

Lamb, H. H. (1964). The role of atmosphere and oceans in relation to climatic changes and the growth of ice sheets on land, *in* "Problems in Paleoclimatology, pp. 332–348. Wiley (Interscience), New York.

Lamb, H. H. (1966). "The Changing Climate." Methuen, London.

Lamb, H. H. (1970). Volcanic dust in the atmosphere . . ., *Trans. R. Philos. Soc. London Math. Phys. Sci.* **266**(1178), 425–533.

Lamb, H. H. (1972). "Climate: Present, Past and Future," Vol. 1. Methuen, London.

Lamb, H. H. (1973). Climate change and foresight in agriculture. *Outlook on Agriculture* **7**(5), 203–210.

Lamb, H. H. (1974). The Current Trend of World Climate, p. 28. Norwich Univ., England.

Lamb, H. H. (1977). "Climate: Present, Past and Future," Vol. 2. Methuen, London.

Lamb, H. H., Malmberg, S. A., and Colebrock, J. M. (1975). Climatic reversal in northern North Atlantic, *Nature (London)* **256**(5517), 479.

Landsberg, H. E. (1956). The climate of towns, "Man's role in changing the face of the Earth," pp. 584–603. Univ. of Chicago Press, Chicago, Illinois.

Landsberg, H. E. (1960). Note on the recent climatic fluctuation in the United States, *J. Geophys. Res.* **65**(5), 1519–1525.

Landsberg, H. E. (1970). Man-made climatic changes, *Science* **170**, 1265–1274.

Landsberg, H. E. (1976). Concerning possible effects of air pollution on climate, *Bull. Am. Meteorol. Soc.* **57**(2), 213–215.

Landsberg, H. E. (1977). The climatic problem, Paper presented at the Leningrad Symposium.

Landsberg, H. E. (1979). The effect of man's activities on climate, *in* "Food, Climate and Man," pp. 187–236. Wiley, New York.

Landsberg, H. E., Groveman, B. S., and Hakkarinen, I. M. (1978). A simple method for approximating the annual temperature of the Northern Hemisphere, *Geophys. Res. Lett.* **5**(6), 505–506.

Landsberg, H. E., and Maisel, T. N. (1972). Micrometeorological observations in an area of urban growth, *Boundary Layer Meteorol.* **2**, 365–370.

Lavrenko, E. M. (1949). On the phytogeosphere, *Vopr.Geogr.* **15**, 53–67.   (R)

Lee, P. S., and Snell, F. M. (1977). An annual zonally averaged global climatic model with diffuse cloudiness feedback, *J. Atmos. Sci.* **34**(6), 847–853.

Lerman, A. *et al.* (1977). Fossil fuel burning: its effects on the biosphere and biogeochemical cycles, *in* "Global Chemical Cycles and their Alterations by Man," pp. 275–289. Dahlem Konferenzen, Berlin.

Lettau, H. A. (1954). Study of the mass, momentum and energy budget of the atmosphere, *Arch. Meteorol. Geoph. Biokl. Ser. A.* **7.**

Lian, M. S., and Cess, R. D. (1977). Energy balance climate models: A reappraisal of ice–albedo feedback, *J. Atmos. Sci.* **34**(7), 1058–1062.

Lieth, H. (1964–1965). Versuch einer kartographischen Darstellung der Productivität der Planzendecke auf der Erde, *Wiesbaden Geogr. Taschenbuch.*

Lindzen, R. S., and Farrell, B. (1977). Some realistic modifications of simple climate models, *J. Atmos. Sci.* **34**(10), 1487–1501.

Lorenz, E. N. (1968). Climatic determinism, *Meteorol. Monogr.* **8**(30), 1–3.

Lyell, Ch. (1830). "Principles of Geology," Vol. 1. London.

Lyell, Ch. (1832). "Principles of Geology," Vol. 2. London.

Lyell, Ch. (1833). "Principles of Geology," Vol. 3. London.

Lyapin, S. E. (1977a). On the simplified model of forming the atmospheric thermal regime, *Proc. State Hydrological Inst.* No. 247, 65–71.   (R)

Lyapin, S. E. (1977b). Some estimates of meteorological solar constant effects on atmospheric thermal regime, *Proc. State Hydrological Inst.* No. 247, 72–79.   (R)

MacAlester, A. L. (1970). Animal extinction, oxygen consumption and atmospheric history, *J. Palaeontol.* **44**(3), 405–415.

MacCormic, R. A., and Ludwig, J. H. (1967). Climate modification by atmospheric aerosols, *Science* **156**, 1358–1359.

Machta, L. (1973). Prediction of CO_2 in the atmosphere, *in* "Carbon and the Biosphere," pp. 21–31. U.S. Atomic Energy Commission, Washington, D.C.

Machta, L. (1978). Energy and carbon dioxide, *in Conf. Climate Energy, Climatol. Aspects Ind. Operat.* pp. 133–139. American Meteorology Society, Boston, Massachusetts.

Manabe, S. (1970). The dependence of atmospheric temperature on the concentration of carbon dioxide, *in* "Global Effects of Environmental Pollution." Reidel, Dordrecht, Holland.

Manabe, S. (1976). Influence of the cloud–radiation feedback mechanism upon the sensitivity of climate, paper presented at the *Tashkent Symp.*

Manabe, S., and Bryan, K. (1969). Climate calculations with a combined ocean–atmosphere model, *J. Atmos. Sci.* **26**(4), 786–789.

Manabe, S., and Hahn, D. G. (1977). Simulation of the tropical climate of an ice age, *J. Geophys. Res.* **82**(27), 3889–3911.

Manabe S., and Wetherald, R. T. (1967). Thermal equilibrium of the atmosphere with a given distribution of relative humidity, *J. Atmos. Sci.* **24**(3), 241–259.

Manabe, S., and Wetherald, R. T. (1975). The effects of doubling the CO_2 concentration on the climate of a general circulation model, *J. Atmos. Sci.* **32**(1), 3–15.

Manabe, S., and Wetherald, R. T. (1980). On the horizontal distribution of climate change resulting from an increase in CO_2 content of the atmosphere, *J. Atmos. Sci.* **37**(1), 99–118.

Marchetti, C. (1976). On geoengineering and the CO_2 problem, Research Memorandum RM-76-17. IIASA, Luxemburg.

Marchuk, G. I. (1974). "Numerical Solution of the Problem of Atmosphere–Ocean Dynamics." Gidrometeoizdat, Leningrad.   (R)

Marchuk, G. I. (1979). Modelling climatic changes and the problem of long-term weather forecasting, *Meteorol. Gidrol.* **7**, 26–36.   (R)

Marchuk, G. I. *et al.* (1977). Hydrodynamical model of atmospheric general circulation, Parts I and II. Preprint CC SD Acad. Sci., USSR, No. 66, 67. Novosibirsk.   (R)

Markov, K. K. (1955). "Essays on the Geography of the Quaternary Period." Geografgiz, Moscow. (R)

Markov, K. K. (1960). "Paleogeography," 2nd ed. Izd. MGU, Moscow. (R)

Martin, R. S. (1967). Prehistoric over-kill. Pleistocene extinctions, *Proc. Cong. Int. Assoc. Q. Res. 3rd* pp. 75–120. Yale Univ. Press, Hartford, Connecticut.

Meleshko, V. P., Shneerov, B. E., Shvets, M. E., and Dmitrieva-Arrago, L. R. (1979). Hydrodynamical three-dimensional model of atmospheric general circulation, *Meteorol. Gidrol.* (6), 21–32. (R)

Menzhulin, G. V. (1976). Climatic change effects on crop productivity, *Proc. Main Geophys. Observ.* No. 365, 41–48. (R)

Mercer, J. H. (1978). West Antarctic ice sheet and CO_2 greenhouse effect: A threat of disaster, *Nature (London)* **271**, 321–325.

Milankovich, M. (1920). "Théorie Mathématique des Phénomènes Thermiques Produits par la Radiation Solaire." Paris.

Milankovich, M. (1930). Mathematische Klimalehre und astronomische Theory der Klimaschwankungen, in "Handbook Klimat." Vol. I. A. Borntrager, Berlin.

Milankovich, M. (1941). Kanon der Erdbestralung und seine Anwendung am Eiszeitenproblem, *R. Serbian Acad.* **132**, 484.

Miles, M. K., and Gildersleeves, P. B. (1977). A statistical study of the likely causative factors in the climatic fluctuations of the last 100 years, *Meteorol. Mag.* **106**, 314–322.

Mintz, Y. (1965). Very long-term global integration of the primitive equations of atmospheric motion, *WMO Tech. Note* No. 66, 141–167. 1965.

Mitchell, J. M., Jr. (1961). Recent secular changes of global temperature, *Ann. N. Y. Acad. Sci.* **95**(1), 235–250.

Mitchell, J. M. (1963). On the world-wide pattern of secular temperature change, *UNESCO, Arid Zone Res.* **20**, 161–180.

Mitchell, J. M. (1968). Concluding remarks. "Causes of climatic changes," pp. 155–159, INQUA.

Mitchell, J. M., Jr. (1971a). Air pollution and climatic change, *Ann. Meeting, 64th, Nov. 28 1970,* American Institute Chemical Engineers, San Francisco, California.

Mitchell, J. M., Jr. (1971b). The effect of atmospheric aerosols on climate with special reference to temperature near the Earth's surface, *J. Appl. Meteorol.* **10**(4), 703–714.

Möller, F. (1963). On the influence of changes in the CO_2 concentration in air on the radiation balance at the Earth's surface and on climate, *J. Geophys. Res.* **68**(13), 3877–3886.

Mokhov, I. I., and Golitsyn, G. S. (1978). Variational estimate of climatic system stability in simple models, *Izv. Akad. Nauk USSR Fis. Atmos. Ocean* (6), 597–606. (R)

National Academy of Science (1977). "Energy and Climate. Studies in geophysics." National Academy of Science, Washington, D.C.

Newell, R. E. (1971). The global circulation of atmospheric pollutants, *Sci. Am.* **224**(1), 32–43.

Newman, J., and Cohen, A. (1972). Climatic effects of aerosol layers in relation to solar radiation, *J. Appl. Meteorol.* **11**(4), 651–657.

Nordhaus, W. D. (1977). Strategies for control of carbon dioxide, Cowles Foundation Discussion Paper. No. 443.

North, G. (1975a). Analytical solution to a simple climate model with diffusive heat transport, *J. Atmos. Sci.* **32**(7), 1301–1307.

North, G. (1975b). Theory of energy-balance climate models, *J. Atmos. Sci.* **32**(11), 2033–2043.

North, G. R., and Coakley, G. A. (1978). Simple seasonal climate models, *Meteorol. Gidrol.* (5), 26–32. (R)

North, G. R. *et al.* (1979). Variational formulation of Budyko–Sellers climate models, *J. Atmos. Sci.* **36**(2), 255–259.

Öpik, E. J. (1953). On the causes of the paleoclimatic variations and of the ice ages in particular, *J. Glacial.* **2**(13), 213–218.

Oerlemans, J., and Van den Dool, H. M. (1978). Energy balance climate models: stability experiments with a refined albedo and updated coefficients for infrared emission, *J. Atmos. Sci.* **35**, 371–381.

Oliver, R. C. (1976). On the response of hemispheric mean temperature to stratospheric dust: an empirical approach, *J. Appl. Meteorol.* **15**(9), 333–350.

Olson, J. S., Pfuderer, H.A., and Chan, Y.-H. (1978). Changes in the Global Carbon Cycle and the Biosphere. Oak Ridge National Laboratory, ORNL/EIS-10.

Oort, A. H., and vonder Haar, T. H. (1976). On the observed annual cycle in the ocean–atmosphere heat balance over the Northern Hemisphere, *J. Phys. Oceanogr.* **6**(6), 781–800.

Owen, T., Cess, R., and Ramanathan, V. (1979). Enhanced CO_2 greenhouse to compensate for reduced solar luminosity on early Earth. *Nature (London)* **277**, 640–642.

Painting, D. J. (1977). A study of some aspects of the climate in the Northern Hemisphere in recent years, Meteorology Office, Sci. Paper No. 35. Her Majesty's Stationery Office, London.

Paltridge, G. W. (1974). Global cloud cover and earth surface temperature, *J. Atmos. Sci.* **31**, 1571–1576.

Pearmen, G. I. (1977). The carbon dioxide–climate problem: recent developments, *Clean Air,* **11**(2), 21–26.

Pivovarova, Z. I. (1968). Long-term variations in direct solar radiation intensity from actinometric observations, *Proc. Main Geophys. Observ.* No. 233, 17–37. (R)

Pivovarova, Z. I. (1977a). "Radiation Characteristics of the USSR Climate." Gidrometeoizdat, Leningrad. (R)

Pivovarova, Z. I. (1977b). The use of ground actimonetric observation data for studying atmospheric transparency, *Meteorol. Gidrol.* (9), 24–31. (R)

Plass, G. N. (1956). The carbon dioxide theory of climatic change, *Tellus* **8**(2), 140–154.

Pokrovskaya, T. V. (1971). On the radiative factors of climate variability, *Meteorol. Gidrol.* (10), 31–38. (R)

Pollack, J. B., Toon, O. B., and Sagan, C. (1975). The effect of volcanic activity on climate, *Proc. WMO/IAMAP Symp. Long-Term Clim. Fluctuat.* pp. 279–285. WMO No. 421, Geneva.

Pollack, J. B. *et al.* (1976). Volcanic explosions and climatic change: A theoretical assessment, *J. Geophys. Res.* **81**(6), 1071–1083.

Pollard, D. (1978). An investigation of the astronomical theory of the ice ages using a simple climate-ice sheet model, *Nature (London)* **272**, 233–235.

Proceedings of Symposium on the Arctic Heat Budget and Atmospheric Circulation (1966). Rand Corp., Santa Monica, California.

Rabinovich, E. (1951). "Photosynthesis and Related Processes," Vol. 2. Wiley (Interscience), New York.

Rakipova, L. R. (1962). Climate changes as a result of the modification of the Arctic ice, *Meteorol. Gidrol.* **9**, 28–30. (R)

Rakipova, L. R. (1966). Changes of the zonal distribution of atmospheric temperature as a result of active climatic modifications, *in* "Contemporary Problems of Climatology," pp. 358–384. Gidrometeoizdat, Leningrad. (R)

Ramanathan, V. (1976). Greenhouse effect due to climatic chlorofluorocarbons: Climatic implications, *Science* **190**, 50–52.

Ramsay, W. (1910). Orogenesis und Klima, *Ofversigt Finska Vetensk Soc. Förh.* **52A,** 140–154.

Raschke, E., Möller, F., and Bandeen, W. (1968). The radiation balance of the earth–atmosphere system over both polar regions obtained from radiation measurements of the Nimbus II meteorological satellite, *Sveriges Meteorol. Hydrol. Inst. Modellanden, Ser. B,* No. 28.

Raschke, E., *et al.* (1973). The radiation balance of the earth–atmosphere system from Nimbus III radiation measurements, NASA Tech. Note D-7249. NASA, Washington, D. C.

Rasool, S. I., and Schneider, S. H. (1971). Atmospheric carbon dioxide and aerosols: effects of large increases on global climate, *Science* **173**(3992), 138–141.

Rauner, Yu. L. (1979). Synchronism of droughts in grain regions of the Northern Hemisphere, *Izv. Akad. Nauk USSR, Ser. Geogr.* (1), 5–19.   (R)

Revelle, R. (1977). Coal and climate, *Sci. News* **112**(5).

Revelle, R., and Munk, W. (1977). The carbon dioxide cycle and the biosphere, *in* "Energy and Climate," pp. 140–158. National Academy of Sciences, Washington, D.C.

Robock, A. (1978). Internally and externally caused climate change, *J. Atmos. Sci.* **35**(6), 1111–1122.

Rodin, L. E., and Basilevich, N. I. (1965). "Dynamics of Organic Matter and Biological Circulation in Basic Vegetation Types." Nauka, Moscow.   (R)

Ronov, A. B. (1959). On the post-Cambrian geochemical history of the atmosphere and hydrosphere, *Geokhimiya* **5,** 397–409.   (R)

Ronov, A. B. (1964). General tendencies in the evolution of the composition of the Earth's crust, ocean and atmosphere, *Geokhimiya* **8,** 715–743.   (R)

Ronov, A. B. (1976). Volcanism, carbon accumulation, life, *Geokhimiya* (12), 1252–1277.   (R)

Ronov, A. B., and Yaroshevsky, A. A. (1976). A model of chemical composition of the Earth's crust, *Geokhimiya* (12), 1763–1794.   (R)

Rubinshtein, E. S. (1946). On the problem of climatic changes, *Proc. NIU, GUGMS* **1**(22), 3–83.   (R)

Rubinshtein, E. S. (1973). "The Structure of Fluctuations of Air Temperature in the Northern Hemisphere." Gidrometeoizdat, Leningrad.   (R)

Rubinshtein, E. S., and Polozova, L. G. (1966). "Contemporary Climatic Changes." Gidrometeoizdat, Leningrad.   (R)

Rukhin, L. B. (1958). Problems of the origin of continental glaciations, *Izv. VGO* **1,** 25–38.   (R)

Rukhin, L. B. (1962). "Principles of General Paleography," 2nd ed. Gostoptekhizdat, Leningrad.   (R)

Rusin, N. P. (1961). "Meteorological and Radiation Regime of the Antarctic." Gidrometeoizdat, Leningrad.   (R)

Rutten, M. G. (1971). "The Origin of Life by Natural Causes." Elsevier, Amsterdam.

Sagan, C. (1975). "The Cosmic Connection." Coronet Books, London.

Sagan, C. (1977). Reducing greenhouses and the temperature history of Earth and Mars, *Nature (London)* **269**(5625), 224–226.

Sagan, C., and Mullen, G. (1972). Earth and Mars: Evolution of atmospheres and surface temperatures, *Science* **177,** 52–56.

Sagan, C., Toon, O. B., and Pollack, J. B. (1978). Human Impact on Climate: Of Global Significance Since the Domestication of Fire (unpublished).

Saks, V. N. (1953). "The Quaternary Period in the Soviet Arctic," 2nd ed. Vodtransizdat, Leningrad and Moscow.   (R)

Saks, V. N. (1960). Geological history of the Arctic Ocean during the Mesozoic era, *Int.*

Geol. Congr. 21st Session, Rep. Sov. Geol. pp. 108–123. Gosgeoltekhizdat, Moscow. (R)

Saltzman, B., and Vernekar, A. D. (1971). Note on the effect of earth orbital radiation variations on climate, *J. Geophys. Res.* **76**(18), 4195–4197.

Sanderson, R. M. (1975). Changes in the area of Arctic sea ice 1966–1974, *Meteorol. Mag.* **104**(1240), 313–323.

Sarasin, P., and Sarasin, F. (1901). La température de la période glaciaire, *Verhandl. Naturforsch. Ges.* **13**, 602–615.

Savinov, S. I. (1913). Maximum values of the solar radiation intensity according to observations in Pavlovsk beginning in 1892, *Izv. Akad. Nauk USSR* **6**(7), 12, 707–720. (R)

Sawyer, J. S. (1963). Notes on the response of the general circulation to changes in the solar constant. Changes of climate, *Proc. Rome Symp., UNESCO* pp. 333–338.

Sawyer, J. S. (1966). Possible variations of the general circulation of the atmosphere. World climate from 8000 to 0 B.C., *Proc. Int. Symp. R. Meteorol. Soc.* pp. 218–229.

SCEP (1970). Man's impact on the global environment, "Report of the Study of Critical Environment Problems." MIT Press, Cambridge, Massachusetts.

Schneider, S. H. (1972). Cloudiness as a global climatic feedback mechanisms, the effects on the radiation balance and surface temperature of variations in cloudiness, *J. Atmos. Sci.* **29**, 1413–1422.

Schneider, S. H. (1975). On the carbon dioxide–climate confusion, *J. Atmos. Sci.* **32**(11), 2060–2066.

Schneider, S. H. (1976). Hearings of the Committee on Science and Technology, U.S. House of Representatives. Washington, D.C.

Schneider, S. H., and Gal-Chen, T. (1973). Numerical experiments in climate stability, *J. Geophys. Res.* **78**(27), 6182–6194.

Schneider, S. H., and Dickinson, R. E. (1974). Climate modeling, *Rev. Geophys. Space Phys.* **12**(3), 447–493.

Schneider, S. H., and Mesirow, L. E. (1976). "The Genesis Strategy." Plenum Press, New York.

Schneider, S. H., and Thompson, S. L. (1978). Ice ages and orbital variations: Some simple theory and modeling, *Quart. Rev.* **12**, 188.

Schopf, T. J. M. *et al.* (1971). Oxygen consumption rates and their palaeoecological significance, *J. Palaeontol.* **45**, 245–252.

Schwarzbach, M. (1950). "Das Klima der Vorzeit. Eine Einführung in die Paläoklimatologie." Stuttgart.

Schwarzbach, M. (1968). Neuere Eiszeithypotesen, *Eiszeitalter Ggw.* **19**, 250–261.

Schwarzschield, M. (1958). "Structure and Evolutions of the Stars." Princeton Univ. Press, Princeton, New Jersey.

Sellers, W. D. (1969). A global climatic model based on the energy balance of the earth–atmosphere system, *J. Appl. Meteorol.* **8**(8), 392–400.

Sellers, W. D. (1970). Comments on "A global climatic model based on the energy balance of the earth–atmosphere system," *J. Appl. Meteorol.* **9**(2), 311.

Sellers, W. D. (1973). A new global climatic model, *J. Appl. Meteorol.* **12**, 241–254.

Sergin, V. Ya., and Sergin, S. Ya. (1969). Studies on climatic change dynamics in the Pleistocene, *Rep. Akad. Nauk USSR* **186**(4), 897–900. (R)

Sergin, V. Ya., and Sergin, S. Ya. (1978). "Systems Analysis of the Problem of Drastic Climatic Change and Glaciations of the Earth." Gidrometeoizdat, Leningrad. (R)

Sharaf, Sh. G., and Budnikova, N. A. (1969).Secular variations of the elements of the Earth's orbit and the astronomical theory of climatic changes, *Proc. Inst. Theoret. Astron.* **14**, 48–85. (R)

Shifrin, K. S., and Minin, I. N. (1957). On the theory of nonhorizontal visibility, *Proc. Main Geophys. Observ.* No. 68, 5–75. (R)

Shifrin, K. S., and Pyatovskaya, N. P. (1959). "Tables of Visibility Range and Daylight Sky Brightness." Gidrometeoizdat, Leningrad. (R)

Shklovksy, I. G. (1973). "The Universe, Life, Reason," 3rd. ed. Nauka, Moscow. (R)

Shnitnikov, A. V. (1969). "Intrasecular Variability of the General Humidity Components," Nauka, Moscow. (R)

Shuleikin, V. V. (1941, 1953, 1968). "Physics of the Sea." Izd. Akad Nauk USSR, Moscow. (1st, 2nd, 3rd eds.) (R)

Shvets, M. E. *et al.* (1970). Numerical model of general atmospheric circulation over a hemisphere, *Proc. Main Geophys. Observ.* No. 256, 3–44. (R)

Simpson, G. (1944). "Tempo and Mode of Evolution." Columbia Univ. Press, New York.

Simpson, G. (1969). "Biology and Man." Harcourt, New York.

Sinitsyn, V. M. (1965). "Ancient Climates in Eurasia, Part I: Paleogene and Neogene." Izd. LGU, Leningrad. (R)

Sinitsyn, V. M. (1967). "Introduction in Palaeoclimatology." Nedra, Leningrad. (R)

Smagorinsky, J. (1963). General circulation experiments with the primitive equations 1. The basic experiment, *Mon. Weather Rev.* **93**, 99–164.

Smagorinsky, J. (1974). Global Atmospheric modeling and the numerical simulation of climate, *in* "Weather and Climate Modification." pp. 633–686. Wiley, New York.

SMIC (1971). "Inadvertent Climate Modification." MIT Press, Cambridge, Massachusetts.

Sochava, A. V., and Glikman, L. S. (1973). Cyclic variations in free oxygen content in the atmosphere and evolution, *in* "Materials of Evolutional Seminar," Part I, pp. 68–87. Vladivostok. (R)

Spirina, L. P. (1971). On the influence of volcanic dust on the thermal regime of the Northern Hemisphere, *Meteorol. Gidrol.* (10), 38–45. (R)

Stuiver, M. (1978). Atmospheric carbon dioxide and carbon reservoir changes, *Science,* **199**(4326), 253–258.

Stumm, W. (1977). Introduction, *in* "Global Chemical Cycles and Their Alterations by Man," pp. 13–21. Dahlem Konferenzen, Berlin.

Su, C. H., and Hsieh, D. Y. (1976). Stability of the Budyko climate model, *J. Atmos. Sci.* **33**(12), 2273–2275.

Suarez, M. J., and Held, I. M. (1976). Modeling climatic response to orbital parameter variations, *Nature (London)* **263**, 46–47.

Teich, M. (1971). Der Verlauf der Jahresmitteltemperature in nordatlantischeuropäischen Raum in den Jahren 1951–1970, *Meteorol. Rundsch.* **24**(5), 137–148.

Temkin, R. L., and Snell, F. M. (1976). An annual zonally averaged hemispherical climatic model with diffuse cloudiness feedback, *J. Atmos. Sci.* **33**, 1671–1685.

Temkin, R. L., Weare, B. C., and Snell, F. M. (1975). Feedback coupling of absorbed solar radiation by three model atmospheres with clouds, *J. Atmos. Sci.* **32**(5), 873–880.

Thomas, R. H., Sanderson, T. J. O., and Rose, K. E. (1979). Effect of climatic warming on the West Antarctic ice sheet, *Nature (London)* **277**, 355–358.

Tikhonov A. N. *et al.* (1969). On the evolution of melting zones in the thermal history of the Earth, *Rep. Akad. Nauk USSR* **188**(2). (R)

Tyndall, J. (1861). On the absorption and radiation of heat by gases and vapours and on the physical connection of radiation absorption and conduction, *Philos. Mag.* **22**(144), 167–194, 273–285.

Van Loon, H., and Williams, J. (1976a). The connection between trends of mean temperature and circulation at the surface: Part 1, Winter; *Mon. Weather Rev.* **104**(4), 365–380.

Van Loon, H., and Williams, J. (1976b). Part 2. Summer, *Mon. Weather Rev.* **104**(8), 1003–1011.

Van Loon, H., and Williams, J. (1976c). Part 3. Spring and Autumn, *Mon. Weather Rev.* **104**(12), 1591–1596.

Van Loon, H., and Williams, J. (1977). Part 4. Comparison of the surface changes in the Northern Hemisphere with the upper air and with the Antarctic in winter, *Mon. Weather Rev.* **105**(5), 636–647.

Velitchko, A. A. (1973). "Natural Process in the Pleistocene." Nauka, Moscow. (R)

Vernekar, A. D. (1972). Long-period global variations of incoming solar radiation, *Meteorol. Monogr.* **12**(34).

Vinnikov, K. Ya. (1965). Outgoing emission of the Earth–atmosphere system, *Proc. Main Geophys. Observ.* No. 168, 123–140. (R)

Vinnikov, K. Ya., and Groisman, P. Ya. (1979). An empirical model of present-day climate change, *Meteorol. Gidrol.* **3**, 25–36. (R)

Voeikov, A. I. (1884). "Climates of the world, particularly of Russia." St. Petersburg. (R)

vonder Haar, T. H., and Suomi, V. E. (1969). Satellite observations of the Earth's radiation budget, *Science* (163), 667–669.

vonder Haar, T. H., and Suomi, V. E. (1971). Measurements of the earth radiation budget from satellites during a five-year period, Part 1, *J. Atmos. Sci.* **28**, 305–314.

Walsh, J. E. (1977). The incorporation of ice station data into a study of recent Arctic temperature fluctuations, *Mon. Weather Rev.* **105**(12), 1527–1535.

Wang, W. C. *et al.* (1976). Greenhouse effects due to man-made perturbations of trace gases, *Science* **194**, 685–690.

Weare B., and Snell, F. M. (1974). A diffuse thin cloud atmospheric structure as a feedback mechanism in global climate modeling, *J. Atmos. Sci.* **31**, 1725–1734.

Wetherald, R. T., and Manabe, S. (1975). The effects of changing the solar constant on the climate of a general circulation model, *J. Atmos. Sci.* **32**(11), 2044–2059.

Wexler, H. (1953). Radiation balance of the Earth as a factor of climate changes, *in* "Climatic Change: Evidence, Causes, and Effects." Harvard Univ. Press, Cambridge, Massachusetts.

Wexler, H. (1956). Variations in insolation, general circulation and climate, *Tellus* **8**, 480–494.

White, O. R. (1967). Sun, *in* "The Encyclopaedia of Atmospheric Sciences and Astrogeology" (R. W. Fairbridge, ed.). Van Nostand-Reinhold, Princeton, New Jersey.

Willett, H. B. (1974). Do recent climatic fluctuations portend an imminent ice age? *Geofis. Int.* **14**(4), 264–295.

Williams, J., Barry, R. G., and Washington, W. M. (1973). Simulation of the climate at the last glacial maximum using the NCAR global circulation model, Inst. Arctic and Alpine Res. Occas. Paper No. 5.

Wilson, A. T. (1964). Origin of ice ages: an ice shelf theory for pleistocene glaciations, *Nature (London)* **201**, 147–149.

Wilson, A. T. (1978). Pioneer agriculture explosion and CO_2 levels in the atmosphere, *Nature (London)* **273**, 40–41.

WMO (1976). Declaration on climate change. *Bull. WMO* **25**, 257–259. (R)

WMO (1977). Technical report of a group of experts of the WMO Executive Committee on climatic change, *Bull. WMO* **26**, 61–69. (R)

Woodwell, G. M. (1978). The carbon dioxide question, *Sci. Am.* **238**(1), 34–43.

Yamamoto, G., and Tanaka, M. (1972). Increase of global albedo due to air pollution, *J. Atmos. Sci.* **29**(8), 1405–1412.

Yudin, M. I. (1950). The influence of forest strips on the turbulent exchange, and their optimal width, *in* "Problems of Hydrometeorological Efficiency of Afforestation for Field Protection." Gidrometeoizdat, Leningrad. (R)

Yung, Y. L. *et al.* (1976). Greenhouse effect due to atmospheric nitrous oxide, *Geophys. Res. Lett.* **3**, 619.

Zakharov, V. F., and Strokina, L. A. (1978). Present-day climatic changes of ice cover of the Arctic Ocean, *Meteorol. Gidrol.* (7), 35–43. (R)

Zeuner, F. (1959). "The Pleistocene Period." Hutchinson, London.

Index

305

International Geophysics Series

EDITED BY

J. VAN MIEGHEM
(July 1959–July 1976)

ANTON L. HALES
(January 1972–December 1979)

WILLIAM L. DONN
Lamont-Doherty Geological Observatory
Columbia University
Palisades, New York